Race Car Design

Derek Seward

Emeritus Professor of Engineering Design
Department of Engineering
Lancaster University

BLOOMSBURY ACADEMIC
LONDON · NEW YORK · OXFORD · NEW DELHI · SYDNEY

BLOOMSBURY ACADEMIC
Bloomsbury Publishing Plc
50 Bedford Square, London, WC1B 3DP, UK
1385 Broadway, New York, NY 10018, USA
29 Earlsfort Terrace, Dublin 2, Ireland

BLOOMSBURY, BLOOMSBURY ACADEMIC and the Diana logo
are trademarks of Bloomsbury Publishing Plc

First published 2014 by PALGRAVE
Reprinted by Bloomsbury Academic, 2022, 2023

Copyright © David Seward 2014

David Seward has asserted his right under the Copyright,
Designs and Patents Act, 1988, to be identified as Author of this work.

For legal purposes the Acknowledgements on p. viii constitute
an extension of this copyright page.

All rights reserved. No part of this publication may be reproduced or
transmitted in any form or by any means, electronic or mechanical,
including photocopying, recording, or any information storage or retrieval
system, without prior permission in writing from the publishers.

Bloomsbury Publishing Plc does not have any control over, or responsibility for,
any third-party websites referred to or in this book. All internet addresses given
in this book were correct at the time of going to press. The author and publisher
regret any inconvenience caused if addresses have changed or sites have
ceased to exist, but can accept no responsibility for any such changes.

A catalogue record for this book is available from the British Library.

A catalogue record for this book is available from the Library of Congress.

ISBN: PB: 978-1-1370-3014-6
ePDF: 978-1-1370-3015-3

Printed and bound in Great Britain

To find out more about our authors and books visit
www.bloomsbury.com and sign up for our newsletters.

Contents

Preface		iv
Symbols		vi
Acknowledgements		viii
1	Racing car basics	1
2	Chassis structure	33
3	Suspension links	61
4	Springs, dampers and anti-roll	90
5	Tyres and balance	119
6	Front wheel assembly and steering	154
7	Rear wheel assembly and power transmission	176
8	Brakes	193
9	Aerodynamics	201
10	Engine systems	227
11	Set-up and testing	241
Appendix 1: Deriving Pacejka tyre coefficients		250
Appendix 2: Tube properties		262
Glossary of automotive terms		265
References		269
Index		271

Preface

The aim of this book is to explain the fundamentals of racing car design using the basic principles of engineering science and elementary mathematics. There is already an extensive list of books that purport to explain this topic. However, with a few honourable exceptions, they tend to fall into one of two camps: either they deal with a highly theoretical and narrow aspect of the subject in a very mathematical way and contain little practical design guidance, or they are written by enthusiastic drivers or constructors who resort to formulas or 'rules of thumb' gained through experience, without proper explanation of the underlying theory. Hopefully this book avoids both of these pitfalls by aiming at a deeper understanding of the principles and avoidance of the 'black art' approach to design. That is not to say that we fully understand every aspect of topics such as tyre/road interaction or aerodynamics. Theory can only take us so far. Even the best designs will require optimisation on the track or in the wind tunnel where the car is tuned to meet the detailed requirements of the specific circuit, driver, tyre compound and weather conditions. The objective of the initial design is therefore to produce a robust solution which is close enough to optimum, so that it can be readily tuned to a wide range of specific conditions.

This book is intended for students on motorsport degree courses, those involved in Formula Student/FSAE and practising car designers and constructors. It will also be of interest to racing drivers and the general reader who is interested in understanding why racing cars are the way they are and why they perform so much better than normal cars on the track. The book is based on the principles of engineering science, physics and mathematics, and hence some previous knowledge of these subjects is required, but only at a relatively elementary level. The book covers the design of most elements of a car including the chassis frame, suspension, steering, brakes, transmission, lubrication and fuel systems; however the internal components of such elements as the engine, gearbox and differential are beyond the scope of this short text. Where relevant, emphasis is placed on the important role that computer tools play in the modern design process.

In many ways the design process for a racing car is much simpler than that for a conventional passenger car because the racing car has a highly focused mission – to propel a driver around a circuit in the shortest time possible. (A passenger car, on the other hand, has a wider remit. It must cope with a varying range of loads from people and luggage, be easy and safe to drive and be comfortable and economical.) The narrow focus of the racing car enables the designer to concentrate almost exclusively on performance issues. The racing car design process can be described as a highly multi-variable problem and inevitably the solution of such problems involves compromise and trade-offs between competing objectives. Resolving these design conflicts presents the skilled designer with the greatest challenges (and pleasure).

The companion website link to this book can be found at

www.palgrave.com/companion/Seward-Race-Car-Design

DEREK SEWARD

Note on glossary terms: The first main use of each glossary term is shown in **bold, *italic*** typeface in the text. The glossary is located on pages 265–268.

Note on 'plates': The 'plates' referred to throughout the text are located between pages 54 and 55.

Online resources to accompany this title are available at: https://www.bloomsburyonlineresources.com/race-car-design. If you experience any problems, please contact Bloomsbury at: onlineresources@bloomsbury.com

Symbols

This list does not include Pacejka tyre model symbols which are defned in the text.

A cross-sectional area (mm^2); amplitude (mm)
A_m brake master-cylinder piston area (mm^2)
A_s brake slave-cylinder piston area (total on one side) (mm^2)
a acceleration (m/s^2)
C damping coefficient; roll couple (Nm)
C_{crit} critical damping coefficient
C_D drag coefficient
C_L lift coefficient
C_0 bearing basic load rating (kN)
C_r bearing dynamic load rating (kN)
D diameter (mm); downforce (N)
E modulus of elasticity (N/mm^2)
F force (N)
F_ϕ lateral load transfer at wheel from roll couple (N)
f frequency (Hz)
f_s sprung mass natural frequency (Hz)
f_u unsprung mass natural frequency (Hz)
G maximum number of g forces; shear modulus (N/mm^2)
g acceleration due to gravity = 9.81 m/s^2
H horizontal component of force (N)
h height (mm)
h_a distance from sprung mass to roll axis (mm)
I second moment of area (mm^4)
K_R suspension ride rate (N/mm)
K_T tyre stiffness (N/mm)
K_W wheel centre stiffness rate (N/mm)
L wheelbase (mm)
l length (mm)
M moment or couple (Nmm)
M_R roll couple (Nmm)
m mass (kg)
P power (W)
P_i absolute pressure (N/mm^2)
P_e Euler buckling load (N)

Symbols

P_m	bearing mean equivalent dynamic load (kN)
P_0	maximum radial load on bearing (kN)
R	radius of curve (m)
R_m	motion ratio
R_R	rolling radius of tyre (mm)
Re	Reynold's number
r_b	brake pad radius (mm)
s	distance travelled (m)
s_0	bearing static safety factor
T	wheel track width (mm); torque (Nm)
T_i	absolute temperature (°C)
t	pneumatic trail (mm); time (s)
u	initial velocity (mm/s)
V_i	volume (m^3)
v	velocity (m/s)
W	weight or wheel loads (N)
Z	elastic section modulus (mm^3)
α	tyre slip angle (deg or rad)
δ	displacement (mm)
$\delta\phi$	wheel displacement from roll (mm)
ζ	damping ratio
θ	angle (deg)
θ_ϕ	roll angle (rad)
μ	coefficient of friction; viscosity (Pa sec)
ρ	density (kg/m^3)
ϕ	wheel camber angle (deg)

Acknowledgements

The publisher and author would like to thank the organisations and people listed below for permission to reproduce material from their publications:

- Avon Tyres Motorsport for permission to reproduce the graphs in Figures 5.8a, 5.8b, A1.1 and A1.4 (adapted by the author).
- Avon Tyres Motorsport for permission to reproduce the data in Tables 5.1 and 5.2 and the 'Avon column' in Table A1.3.
- Caterham F1 Team for permission to reproduce the photographs in Figures 3.20, 5.17 and 9.1.
- Mike Pilbeam for permission to reproduce the photograph in Figure 2.11, which was taken by Rick Wilson of Redline Design.

The publisher and author would like to acknowledge the companies listed below for the use of their software:

- Figure 10.6 uses software under licence from DTAfast.
- Figures 1.2 and 1.21 use software under licence from ETB Instruments Ltd – DigiTools Software.
- Figures 2.3a–c, 2.6a–b, 2.9, 2.14, 2.15a–c, 3.19 and Plates 1, 2, 3 and 5 use software under licence from LISA.
- Figure 10.3 uses software under licence from Lotus Engineering, Norfolk, England.
- Figures 1.3, 5.15, 5.16, Table 6.1, Figures 7.2, 7.4, 7.5, A1.2 and A1.5 use Excel® under licence from Microsoft®.

The following figures use Visio® under licence from Microsoft®:

Figures 1.1, 1.2, 1.4, 1.5, 1.6, 1.8, 1.9, 1.10, 1.11, 1.12, 1.13, 1.14, 1.15, 1.16, 1.17, 1.18, 1.19, 1.20, 1.21, 1.22, 1.23, 1.24, 2.1a–c, 2.2, 2.3a–c, 2.4, 2.5, 2.6a–b, 2.7, 2.8, 2.9, 2.10, 2.11, 2.12a–b, 2.13a–b, 2.17, 2.18, 3.2, 3.5, 3.8a–c, 3.9, 3.10, 3.11, 3.12, 3.13, 4.3, 4.4, 4.7, 4.8, 4.9a–c, 4.10, 4.13, 5.1a–d, 5.2a–b, 5.3, 5.4, 5.5, 5.6, 5.7, 5.10, 5.11, 5.12, 5.13, 5.14, 5.18, 6.2, 6.4, 6.5, 6.6, 6.7, 6.8, 7.3, 7.6, 8.3, 9.2, 9.3, 9.4, 9.5, 9.6a–b, 9.7, 9.8, 9.9, 9.10, 10.2, 10.4, 10.7, 10.8, 10.9 and 11.1.

Acknowledgements

The following figures use software under licence from SketchUp:

Figures 1.4, 1.5, 1.6, 1.8, 1.9, 1.10, 1.12, 1.13, 1.15, 1.16, 1.17, 1.22, 1.23, 2.8, 3.8a–c, 3.9, 3.10, 3.11, 3.12, 3.13, 5.1a–d, 5.2a–b, 6.5, 6.6, 8.3, 9.2, 9.7, 9.8, 9.9, 9.10 and 11.1.
Plates 6 and 7 use software under licence from SolidWorks.

The following figures and plate use software under licence from SusProg:

Figures 3.1a–d, 3.3a–b, 3.4a–c, 3.6a–b, 3.7a–b, 3.14a–d, 3.15, 3.16, 3.17, 3.18, 3.20, 3.21, 3.22, 3.23, 6.9 and Plate 4.

The following figures use ViaCAD software under licence from Punch!CAD:

Figures 2.1a–c, 2.2, 2.3a–c, 2.4, 2.5, 2.7, 2.9, 2.10, 2.11, 2.12a–b, 2.13a–b, 4.5, 4.6, 4.11a–d, 4.12, 5.9, 6.3, 6.10, 6.11, 6.12, 6.13, 6.14a–b, 6.15a–b, 7.8, 7.9, 7.10a–b, 8.1 and 10.1.

1 Racing car basics

LEARNING OUTCOMES

At the end of this chapter:
- You will understand the basic elements of car racing
- You will be able to calculate the varying loads on the wheels of a racing car as it accelerates, brakes and corners, and appreciate how these loads are influenced by aerodynamic downforce
- You will be able to identify some important design objectives for a successful racing car

1.1 Introduction

This chapter introduces many of the key concepts that must be grasped to obtain a good understanding of racing car design. It also contains signposts to later chapters where topics are covered in more depth. By its nature, racing is a highly competitive activity and the job of the designer is to provide the driver with the best possible car that hopefully has a competitive advantage. To do this we need answers to the following questions:

- What does a racing car have to do?
- What is the best basic layout of a car for achieving what it has to do?
- How can the car be optimised to perform better than the competition?
- What loads and stresses is the car subjected to, and how can it be made adequately safe and robust?

This chapter will start to provide some of the answers to these questions.

1.2 The elements of racing

Motor racing can take many forms ranging from short **hill climbs** and **sprints**, where the driver competes against the clock, to conventional head-to-head **circuit racing** such as Formula 1 and IndyCar; however there are common elements to all forms. In general the aim of all racing is to cover a particular piece of road or circuit in the shortest possible time. To do this the driver must do three things:

- Accelerate the car to the fastest possible speed.
- Brake the car as late as possible over the minimum possible distance.
- Go round corners in the minimum time and, more importantly, emerge from corners carrying the maximum possible velocity so that a speed advantage is carried over the ensuing straight.

From the above it can be seen that the competitive driver will spend virtually no time 'cruising' at constant velocity. The only time this will occur is either queuing in traffic or flat-out on a long straight. Also, of course, the skilled driver may combine these basic elements by 'accelerating out of a corner' or 'braking into a corner'.

This is illustrated in *Figures 1.1* and *1.2* which show the layout of a circuit together with a plot of speed data for one lap. The labels indicate matching points on both figures. Note that the slope of the curve in *Figure 1.2* is steeper during braking than during acceleration. This is for three reasons: firstly, at faster speeds the rate of acceleration is limited by the power of the engine; secondly, braking uses the grip from all four wheels whereas, in this case, acceleration uses only rear wheel grip; thirdly, at fast speeds, the car develops significant aerodynamic drag forces which assist braking but impede acceleration.

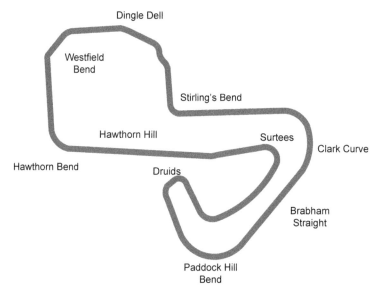

Figure 1.1
Brands Hatch circuit, UK

The three basic elements of racing all involve a form of **acceleration** or change in velocity. In the case of cornering this is **lateral acceleration** and braking can be considered to be **negative acceleration**. We know from Newton's first law of motion that:

> 'An object in motion stays in motion with the same speed and in the same direction unless acted upon by an external force.'

Chapter 1 **Racing car basics**

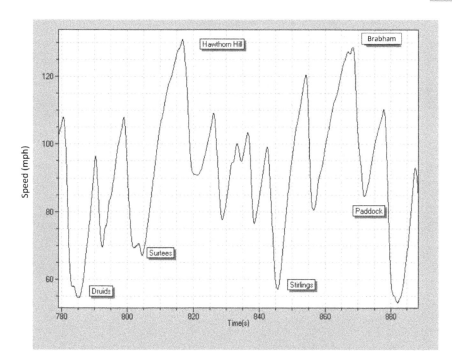

Figure 1.2
Brands Hatch speed data (produced with ETB Instruments Ltd – DigiTools Software)

Consequently in order to accelerate or change direction, the car must be subject to an external force and the principal source of such a force is at the interface between the tyres and the road – known as the **tyre contact patch**. (Clearly external aerodynamic forces also exist and these will be considered later.) Thus it can be concluded that the ability of a car to accelerate, brake and change direction depends upon the frictional force developed between the rubber tyre and the road surface. This force is normally referred to as **traction** or **grip** and its maximisation is an important design criterion for a competitive car.

Classical, or Coulomb, friction has a simple linear relationship between the applied normal load and a constant coefficient of friction, μ (mu):

Friction force = normal load × μ

As we shall see when we look at tyre mechanics in more detail later, the contact patch between a tyre and the road **does not** follow this simple law. *Figure 1.3* shows the relationship between vertical wheel load and maximum lateral grip for a typical racing tyre and compares it to simple Coulomb friction with μ = 1 (dashed line).

We will see later that the lack of linearity (i.e. the coefficient of friction not being constant) provides a powerful means by which a car's handling is tuned for peak performance. It can be concluded from *Figure 1.3* that:

- As the vertical load is increased on the wheel, the grip increases, but at a progressively slower rate. This is known as **tyre sensitivity**.

Figure 1.3
Typical racing tyre grip

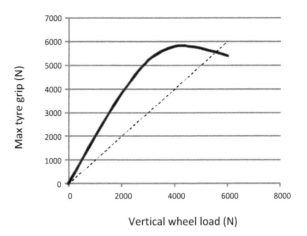

- Eventually the level of grip peaks, and then starts to fall with increasing wheel load. The tyre has become **overloaded**.
- The value of grip divided by vertical wheel load at a specific point in *Figure 1.3* can be considered to be an instantaneous coefficient of friction.

It is clear that knowledge of the normal force at each tyre contact patch – i.e. the individual vertical wheel loads – is vital for many aspects of racing car design. They are used to determine the loads in the chassis, brake components, suspension members, transmission etc., as well as for tuning the fundamental handling and balance of the car. We will look at static wheel loads and then see how they change when the car is subjected to the three elements of racing – braking, acceleration and cornering. First it is necessary to determine the position of the car's **centre of mass** which is often referred to as the **centre of gravity**. The centre of mass is the point where all of the mass can be considered to be concentrated. Knowledge of its location is important to car designers as this determines the weight distribution between the front and rear wheels. Also the height of the centre of mass above the ground influences the degree to which the car rolls on corners as well as the amount of weight that transfers between the wheels during braking, acceleration and cornering.

1.3 Position of centre of mass of a vehicle

At the preliminary design stage it is necessary to estimate the centre of mass of each major component as it is added to the scheme. The final positional relationship between the components and the wheels can then be adjusted to achieve the desired front/rear weight distribution.

To illustrate the process, *Figure 1.4* shows just a couple of components, together with distances from their individual centres of mass to a common point. In this case the common point is the front contact patch, x.

Chapter 1 Racing car basics

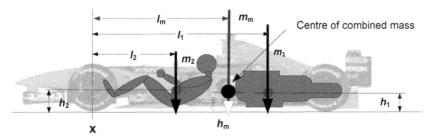

Figure 1.4
Calculating the position of the centre of mass

The magnitude (m) and location (l,h) of the centre of mass of each individual component is either measured or estimated. The objective is to find the value of the combined mass, m_m, and its location relative to the common point, l_m and h_m.

The combined mass is simply the sum of the individual components. For a total of n number components, this is shown mathematically as:

$$m_m = \Sigma(m_1 + m_2 + \ldots m_n) \qquad [1.1]$$

The location of the combined centre of mass is given by:

$$l_m = \frac{\Sigma(m_1 l_1 + m_2 l_2 + \ldots m_n l_n)}{m_m} \qquad [1.2]$$

$$h_m = \frac{\Sigma(m_1 h_1 + m_2 h_2 + \ldots m_n h_n)}{m_m} \qquad [1.3]$$

The above process simply ensures that the combined mass of the components exerts the same moment about the front contact patch as the sum of all the individual components.

EXAMPLE 1.1

The following data is relevant to the two components shown in *Figure 1.4*.

Determine the magnitude and location of the combined centre of mass.

Item	Mass (kg)	Horiz. dist. from x (mm)	Vert. dist. from ground (mm)
Engine	120	2100	245
Driver	75	1080	355

Solution

From *equation [1.1]*

Combined mass, m_m = 120 + 75 = 195 kg

From *equation [1.2]*

Horiz. distance to combined mass, l_m = $\dfrac{(120 \times 2100) + (75 \times 1080)}{195}$ = 1708 mm

From *equation [1.3]*

$$\text{Vertical distance to combined mass, } h_m = \frac{(120 \times 245) + (75 \times 355)}{195} = 287 \text{ mm}$$

Answer: **Combined mass = 195 kg acting 1708 mm horizontally from x and at a height of 287 mm above ground**

Clearly, in the case of a real car, there are many more components to consider and the use of a spreadsheet is desirable. *Table 1.1* shows such a spreadsheet which can be downloaded from www.palgrave.com/companion/Seward-Race-Car-Design for your own use. Your own data can be input into the shaded cells.

Once a car has been constructed the position of the centre of mass should be confirmed by physical measurements and this is discussed in *Chapter 11* as part of the set-up procedure.

1.4 Static wheel loads and front/rear weight balance

The static case refers to the loads on the car when it is **not** being subjected to accelerations from accelerating, braking or cornering. The car should be considered when fully laden with driver and all fluids. These are the loads that would be measured if the car was placed on level ground in the pits. [Up to now we have referred to the **mass** of components in **kilograms**. However the terms 'load' and 'weight' actually imply **force** which is, of course, measured in **Newtons**. Consequently from now on we will consider the **forces, *W*,** on the car where force (N) = mass (kg) × acceleration (m/s²), where, for vertical loads, the acceleration = g (9.81 m/s²).]

Figure 1.5
Calculating static wheel loads

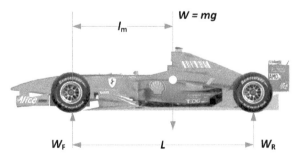

Figure 1.5 shows a car where the magnitude and position of the centre of mass has been determined. We wish to find the static wheel loads. Knowing

Chapter 1 **Racing car basics**

Element	Mass (kg)		Horiz. dist. front axle (mm)	H moment (kgm)	Vert. dist. ground (mm)	V moment (kgm)
Car						
Front wheel assemblies	32.4		0	0	280	9072
Pedal box	5		0	0	260	1300
Steering gear	5		300	1500	150	750
Controls	3		200	600	400	1200
Frame + floor	50		1250	62500	330	16500
Body	15		1500	22500	350	5250
Front wing	5		−450	−2250	90	450
Rear wing	5		2700	13500	450	2250
Fire extinguisher	5		300	1500	260	1300
Engine assembly + oil	85		1830	155550	300	25500
Fuel tank (full)	25		1275	31875	200	5000
Battery	4		1200	4800	120	480
Electrics	4		1500	6000	200	800
Exhaust	5		1750	8750	350	1750
Radiator + water	10		1360	13600	150	1500
Rear wheel assemblies + drive shafts + diff.	58		2300	133400	280	16240
Reversing motor	6		2500	15000	280	1680
Ballast	0		1200	0		0
Other 1				0		0
Other 2				0		0
Other 3				0		0
Other 4				0		0
Other 5				0		0
Total car	**322.4**		**1454**	**468825**	**282**	**91022**
Driver						
Weight of driver	80		Distance front axle to pedal face	50		

	Mass (kg)	Horiz. dist. sole foot (mm)	Horiz. dist. front axle (mm)	H moment (kgm)	Vert. dist. ground (mm)	V moment (kgm)
Feet	2.8	40	90	250	310	859.733333
Calves	7.7	350	400	3072	360	2764.8
Thighs	17.3	760	810	13997	295	5097.6
Torso	36.9	1050	1100	40597	300	11072
Forearms	3.2	800	850	2720	400	1280
Upper arms	5.3	1100	1150	6133	420	2240
Hands	1.3	650	700	896	510	652.8
Head	5.5	1200	1250	6933	670	3716.26667
Total driver	**80**	**5950**	**6350**	**74598.4**	**346**	**27683.2**
Grand total	**402.4**		**1350**	**543423**	**295**	**118705.2**
Rear axle load	236					
Front axle load	166					
Ratio F/R	**41.3%**	**58.7%**				

Table 1.1 Spreadsheet for calculating centre of mass

Race car design

the wheelbase and the horizontal position of the centre of mass we can simply take moments about the front axle to find the rear axle load, W_R:

$$\text{Rear axle load, } W_R = W \times \frac{l_m}{L}$$

From vertical equilibrium:

$$\text{Front axle load, } W_F = W - W_R$$

It should be noted that *Figure 1.5* is a **free-body diagram**. If the car is considered to be floating weightlessly in space the three acting forces, W, W_F and W_R, must keep it in static equilibrium, i.e. the downward force from gravity, W, must be equal and opposite to the sum of the wheel reaction forces, W_F and W_R. This is why the wheel forces are shown upwards. They represent the forces from the road acting on the car. We will make extensive use of free-body diagrams throughout this book.

EXAMPLE 1.2

For the car shown in *Figure 1.6*:

(a) Determine the static axle loads.
(b) Calculate the percentage front/rear distribution.
(c) Estimate individual static wheel loads.

Figure 1.6
Position of centre of mass

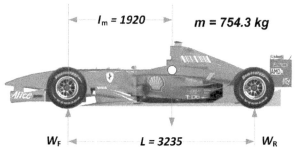

(a) Weight of car, $W = 754.3 \times 9.81 = 7400$ N

Rear static axle load, $W_R = 7400 \times \dfrac{1920}{3235} = 4392$ N

Front static axle load, $W_F = 7400 - 4392 = 3008$ N

(b) % to front $= \dfrac{3008}{7400} \times 100 = 40.6\%$

% to rear $= 100 - 40.6 = 59.4\%$

(c) A circuit racing car is usually expected to have good left/right balance and therefore the individual wheel loads can be assumed to be half of the axle loads. Hence:

Rear static wheel loads, W_{RL} and $W_{RR} = \dfrac{4392}{2} = 2196$ N

Front static wheel loads, W_{FL} and $W_{FR} = \dfrac{3008}{2} = 1504$ N

Answer: Static axle loads = 4392 N rear and 3008 N front
Distribution = 40.6% front and 59.4% rear
Static wheel loads = 2196 N rear and 1504 N front

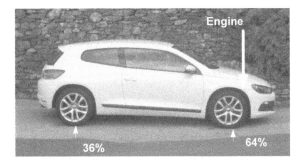

(a) VW Scirocco – front engine, front-wheel drive, equal tyres

(b) BMX 3 series – front/mid engine, rear-wheel drive, equal tyres

(c) Formula car – mid engine, rear-wheel drive, wider rear tyres

(d) Porsche 911 – rear engine, rear-wheel drive, much wider rear tyres

Figures 1.7a – d
Weight distribution as a function of driven wheels, engine position and tyre selection

Clearly the designer can influence the front/rear weight balance by moving certain components, such as the battery or hydraulic pumps. A significant change results from modifying the location of the front and/or rear axles relative to the significant mass of the engine and gearbox. In addition competitive cars are invariably built significantly lighter than the minimum weight specified in their formula technical regulations. The difference is then made up by the addition of heavy ballast which is strategically placed to give the best front/rear balance.

What, then, is the optimum front/rear weight ratio? From a handling point of view it can be argued that a 50:50 ratio is optimum. However, as we shall see shortly, for accelerating off the line, there is a clear advantage in having more weight over the driven wheels. Racing cars typically aim for about a 44:55 front/rear ratio and address the handling issue by means of wider rear tyres. *Figures 1.7a–d* show how different weight distributions have resulted from particular combinations of driven wheels, engine position and tyre selection.

The position of the fuel tank(s) presents a challenge as the weight of fuel clearly varies throughout the race. In Formula 1, where refuelling is no longer allowed, the cars start with up to 170 kg of fuel. The solution is to put the fuel tanks as close to the centre of mass as possible, so that, as it is used, the balance of the car does not change.

The three elements of racing will now be considered in more detail.

1.5 Linear acceleration and longitudinal load transfer

The starting point for understanding linear acceleration is Newton's second law of motion:

> *'The acceleration (a) of an object is directly proportional to the magnitude of the applied force (F) and inversely proportional to the mass of the object (m).'*

This can be written as:

$$a = \frac{F}{m} \quad\quad [1.4]$$

As the mass of a car can be considered constant, the rate of acceleration is dependent upon the force available to propel the car forward. *Figure 1.8* shows this traction force acting at the contact patch of the driven rear wheels. We can convert a dynamic analysis into a simple static analysis by invoking d'Alembert's principle which states that the car effectively 'resists' forward acceleration with an imaginary inertial reaction force that acts through the centre of mass. This is equal and opposite to the traction force and is shown as the resistive force in *Figure 1.8*. The fact that the traction force occurs at road level and the resistive force at the level of the centre of mass means that an out-of-balance couple or moment is set up. This causes changes to the static axle loads, W_F and W_R. The magnitude of the change, ΔW_x, is known as **longitudinal load transfer** and it is added to the static rear axle load and subtracted from the static front axle load. This explains why, when accelerating hard, the front of a car rises and the rear drops – known as **squat**.

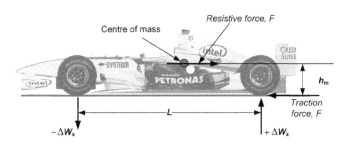

Figure 1.8
Linear acceleration and longitudinal load transfer

By taking moments about the front contact patch:

$$F \times h_m = \Delta W_x \times L$$

∴ Longitudinal weight transfer, $\Delta W_x = \pm \dfrac{F h_m}{L}$ [1.5]

(Incidentally, if the force accelerating the car was applied at the level of the centre of mass, instead of at road level – say by a jet engine, there would be no longitudinal weight transfer, as all the forces act through one point and there is no out-of-balance couple.)

When accelerating a car from the start-line up to its maximum speed, we can consider two distinct stages:

Stage 1 – Traction limited

During initial acceleration off the start-line the value of the traction force, F, is limited by the frictional grip that can be generated by the driven tyres. The problem, at this stage, for the driver, is to avoid wheel-spin.

Stage 2 – Power limited

As the speed of the car increases the point will be reached where the engine cannot provide enough power to spin the wheels, and from this point onwards maximum acceleration is limited by engine power. As speed increases still further, aerodynamic drag force and other losses build until all of the engine power is needed to overcome them. At this point further acceleration is not possible and the car has reached its maximum speed or **terminal velocity**.

Race car design

1.5.1 Traction-limited acceleration

The initial traction-limited stage produces the highest levels of traction force and hence longitudinal load transfer. It is the case that, for design purposes, this produces the highest loads on the rear suspension and transmission. *Figure 1.9* shows the same car with static loads added and the imaginary resistive force removed.

Figure 1.9
Traction-limited acceleration

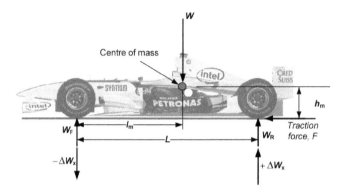

From *equation [1.5]*

$$\text{Longitudinal load transfer, } \Delta W_x = \frac{Fh_m}{L}$$

$$\text{Traction force, } F = (W_R + \Delta W_x) \times \mu$$

$$F = \left(W_R + \frac{Fh_m}{L}\right) \times \mu$$

$$\therefore F - \frac{Fh_m \mu}{L} = W_R \mu$$

$$F\left(1 - \frac{h_m \mu}{L}\right) = W_R \mu$$

$$F = \frac{W_R \mu}{1 - \dfrac{h_m \mu}{L}} \qquad [1.6]$$

At this stage it is necessary to assume a value for the coefficient of friction, μ. It has already been stated that the value for the tyre contact patch is not in fact a constant – see *Figure 1.3* – however an appropriate average value for a warm racing slick tyre is generally assumed to be in the range 1.4 to 1.6. This compares to about 0.9 for an ordinary car tyre.

Once *equation [1.6]* is solved for F, it is an easy matter to substitute back into *equation [1.5]* to obtain the longitudinal weight transfer, ΔW_x. This is demonstrated in the following *Example 1.3*.

EXAMPLE 1.3

For the car shown in *Figure 1.10*:

(a) Estimate the individual wheel loads during maximum acceleration assuming an average coefficient of friction, μ, between the tyre and the road of 1.5.
(b) If the rear tyres have a rolling radius of 275 mm, estimate the peak torque through the transmission when accelerating off-the-line.
(c) Calculate the maximum acceleration in both m/s² and equivalent *g* force.

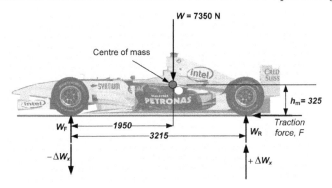

Figure 1.10
Calculating wheel loads during acceleration

(a) Weight of car, W = 7350 N

Static rear axle load, W_R = $7350 \times \dfrac{1950}{3215}$ = 4458 N

Static front axle load, W_F = 7350 − 4458 = 2892 N

From *equation [1.6]*
Traction force, F = $\dfrac{W_R \mu}{1 - \dfrac{h_m \mu}{L}}$ = $\dfrac{4458 \times 1.5}{1 - \dfrac{325 \times 1.5}{3215}}$

= **7882 N**

From *equation [1.5]*
Longitudinal load transfer, $\Delta W_x = \pm \dfrac{F h_m}{L}$ = $\pm 7882 \times \dfrac{325}{3215}$

= **±797 N**

Rear wheel loads, W_{RL} and W_{RR} = $\dfrac{4458 + 797}{2}$ = **2628 N**

Front wheel loads, W_{FL} and W_{FR} = $\dfrac{2892 - 797}{2}$ = **1048 N**

(b) Peak torque at rear wheels, T_{wheels} = $(W_{RL} + W_{RR}) \times \text{rad.} \times \mu$
  = $(2628 + 2628) \times 275 \times 1.5$
  = 2 168 000 Nmm = **2168 Nm**

(c) $$\text{Mass of car} = \frac{7350}{9.81} = 749.2 \text{ kg}$$

From *equation [1.4]*
$$\text{Acceleration, } a = \frac{F}{m} = \frac{7882}{749.2}$$
$$= 10.52 \text{ m/s}^2$$
$$= \frac{10.52}{9.81} = 1.072g$$

Answer: Wheel loads = 2628 N rear and 1048 N front
Torque through transmission = 2168 Nm
Acceleration = 10.52 m/s² = 1.072g

Comment:
The above rear wheel loads and torque represent an important load case for the design of the transmission, rear wheel assemblies and suspension components.

It should be recognised that if a car is to achieve the above peak values of traction force and acceleration then it is necessary for it to have an adequate power-to-weight ratio (rarely a problem for a racing car), suitable transmission gearing and a driver capable of appropriate clutch and throttle control (or an automated traction control system). These issues will be dealt with later in the book.

1.5.2 Power-limited acceleration

When a force, such as the traction force causing acceleration, moves through a distance, it does **work**:

Work = force × distance (Nm or Joules)   [1.7]

Power is the rate of doing work:

$$\text{Power} = \frac{\text{force} \times \text{distance}}{\text{time}}$$

= force × speed (Nm/s or Watts)

or Force = $\frac{\text{power}}{\text{speed}}$ (N)   [1.8]

It can be seen from *equation [1.8]* that, if power is limited, the traction force must reduce as speed increases. In this case it is not appropriate to consider absolute peak engine power as this is generally only available at specific engine revs. The average power available at the wheels as the driver moves through the gears will be a bit less. In addition some power is lost in 'spinning-up' the transmission components and wheels as well as overcoming transmission friction. Furthermore not all of the traction force is available for accelerating the car. Some of it must be used to overcome further losses. The two principal additional losses are:

- rolling resistance from the tyres,
- aerodynamic drag.

Rolling resistance largely results from the energy used to heat the tyre as the rubber tread deforms during rolling. The degree of resistance is related to the vertical load carried by each tyre as well as the rolling velocity. It depends upon the tyre construction, wheel diameter and the road surface, but for racing tyres it can be approximated to 2% of the car weight.

Aerodynamic drag is dependent on the frontal area of the car and the degree of streamlining. It increases with the square of the velocity and hence becomes the dominant loss at high speeds. Aerodynamic forces are dealt with in more detail in *Chapter 9*.

Figure 1.11 shows how the net force available to accelerate the car reduces as speed increases. When this force is zero the car has reached its maximum, or terminal, speed.

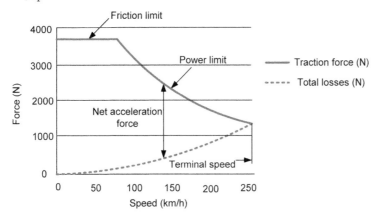

Figure 1.11
The force available for acceleration

Acceleration will be considered again in *Chapter 7* where we will consider the implications of choosing the best gear ratios for peak performance.

1.6 Braking and longitudinal load transfer

As all four wheels are braked, the braking force can be considered to be simply the weight of the car (N) multiplied by an assumed average tyre/ground coefficient of friction, μ:

Braking force, $F = W \times \mu$ [1.9]

Figure 1.12
Braking and longitudinal load transfer

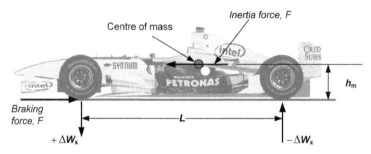

It can be seen from *Figure 1.12* that the forces are reversed compared to the acceleration case shown in *Figure 1.8*. In this case load is transferred from the rear wheels to the front causing the nose to dip – known as **dive**.

From *equation [1.5]*

Longitudinal weight transfer, $\Delta W_x = \pm \dfrac{Fh_m}{L} = \pm \dfrac{W\mu h_m}{L}$ [1.10]

EXAMPLE 1.4

For the same car as *Example 1.3* and shown in *Figure 1.13*:

(a) Estimate the individual wheel loads during maximum braking assuming an average coefficient of friction, μ, between the tyre and the road of 1.5.
(b) Calculate the maximum deceleration in both m/s² and *g* force.

Figure 1.13
Calculating wheel loads during braking

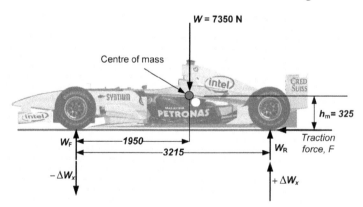

Chapter 1 Racing car basics

(a) Weight of car, W = 7350 N

As before:

Static rear axle load, W_R = $7350 \times \dfrac{1950}{3215}$ = 4458 N

Static front axle load, W_F = $7350 - 4458$ = 2892 N

From *equation [1.9]*

Braking force, $F = W \times \mu$ = 7350×1.5
= **11 025 N**

From *equation [1.5]*

Longitudinal weight transfer, $\Delta W_x = \pm \dfrac{Fh_m}{L}$ = $11\,025 \times \dfrac{325}{3215}$
= **± 1115 N**

Front wheel loads, W_{FL} and W_{FR} = $\dfrac{2893 + 1115}{2}$ = **2004 N (55%)**

Rear wheel loads, W_{RL} and W_{RR} = $\dfrac{4458 - 1115}{2}$ = **1672 N (45%)**

Mass of car = $\dfrac{7350}{9.81}$ = 749.2 kg

(b) From *equation [1.4]*

Deceleration, $a = \dfrac{F}{M}$ = $\dfrac{11\,025}{749.2}$

= **14.72 m/s²**

= $\dfrac{14.72}{9.81}$ = **1.5g**

Answer: Wheel loads = **2004 N front and 1672 N rear**
Deceleration = **14.72 m/s² = 1.5g**

Comments:
It can be seen that
1. *The above wheel loads represent an important load case for the design of the brake system and the front wheel assemblies and suspension components.*
2. *Because braking involves grip from all four wheels and aerodynamic drag, a car decelerates at a higher rate than that achieved during acceleration.*
3. *During maximum braking the front wheel loads (and hence brake forces) are usually greater than the rear which explains why road cars often have bigger brake discs on the front. This is despite the fact that, in this case, the static rear wheel loads are larger than the front.*

Race car design

4. *The deceleration in terms of g force magnitude is equal to the average friction coefficient, μ. (However this is only true if aerodynamic drag and downforce are ignored.)*

Details of brake system design will be considered in more detail in *Chapter 8*.

1.7 Cornering and total lateral load transfer

Cornering can be considered to be the most conceptually challenging element of racing. It is not immediately obvious why a car travelling at constant speed around a corner should be subject to acceleration. The key lies in the fact that **velocity** is a **vector quantity**. A vector has both magnitude and direction (unlike **speed** which is a **scalar quantity** and has only magnitude). Although the magnitude may remain constant, a cornering vehicle is subject to changing direction and hence changing velocity. Changing velocity is acceleration and, because a car has mass, this requires a force – so-called **centripetal force**.

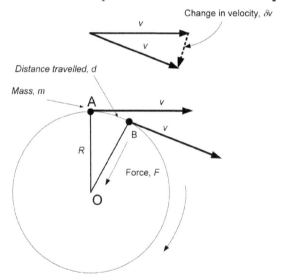

Figure 1.14
Deriving the centripetal force formula

Consider the familiar problem of a mass, m, on the end of a string and being swung in a circle – *Figure 1.14*. In a small increment of time (δt) the mass moves from point A to B (distance exaggerated for clarity). The arrows emanating from A represent the vectors of velocity for points A and B, i.e. they are the same length (magnitude) but the directions are the tangents to the circle at A and B. The dashed arrow indicates the change in velocity (δv). As the increment becomes small the direction of this velocity change vector points to the centre of rotation, O.

Also, as the increment becomes small, the lines from A and those from O form **similar triangles**. Hence:

$$\frac{\delta v}{v} = \frac{d}{R}$$

But Distance travelled, $d = v \times \delta t$

Therefore $$\frac{\delta v}{v} = \frac{v \times \delta t}{R}$$

Divide both sides by δt: $$\frac{\delta v}{v \times \delta t} = \frac{v}{R}$$

Multiply both sides by v: $$\frac{\delta v}{\delta t} = \frac{v^2}{R}$$

Hence $a = \dfrac{v^2}{R}$

For an object with mass, m

Centripetal force, $\mathbf{F} = ma = \dfrac{mv^2}{R}$ [1.11]

Centripetal force is the force that the string exerts on the mass. The equal and opposite force that the mass exerts on the string is the so-called **centrifugal force**, and this will act through the centre of mass.

In the case of a car, the centripetal force is provided by lateral grip from the tyres as shown in *Figure 1.15*. This is often referred to as the **cornering force**. The equal and opposite centrifugal force passes through the centre of mass. The fact that the centre of mass is not, in this case, in the centre of the

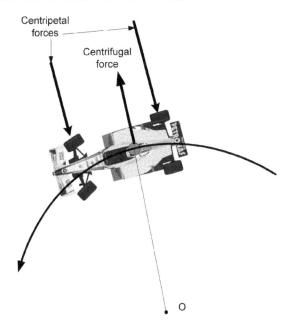

Figure 1.15
Racing car cornering

wheelbase means that the tyre lateral grip forces are unequal. The designer must therefore provide additional grip to the rear wheels in this case – say by using wider tyres. For peak cornering performance the front wheels must give way at roughly the same time as those at the rear. This is what is meant by a **balanced** car. If the front wheels give way before the rears the car is said to **understeer** and the car will refuse to turn and carry straight on at a bend. If the rear wheels give way before the fronts the car is said to **oversteer** and the car is likely to spin. These issues will be considered in much more detail in *Chapter 5* where a more rigorous definition of understeer and oversteer is provided and we will see how calculations can be applied to achieve the necessary balance. However fine-tuning for balance is invariably required by driving on the circuit where adjustments are made to suit the particular driver, tyres, road surface and weather conditions. More is said about this in *Chapter 11*.

It is clearly an easy matter to use the centripetal force *equation [1.11]* to find the required cornering force for a car going round a specific corner at a specific speed. However, the designer is more interested in maximising the cornering force and expressing the cornering performance in terms of the number of lateral *g* forces that the car can attain. As with braking, we can approximate this by estimating the average coefficient of friction, μ, at the tyre contact patch. For a car with no aerodynamic downforce:

Maximum cornering force, $F = W \times \mu$ (N) [1.12]

where W is the weight of the car.

It can be seen from *Figure 1.16* that because the centrifugal force passes through the centre of mass, which is above the road surface, an overturning moment or couple is created which causes lateral load transfer, ΔW_y. When cornering, the load on the outer wheels increases and the load on the inner wheels decreases by the same amount.

Total lateral load transfer, $\Delta W_y = \pm \dfrac{F h_m}{T}$ [1.13]

where T is the distance between the centre of the wheels or **track**.

Figure 1.16
Lateral load transfer during cornering

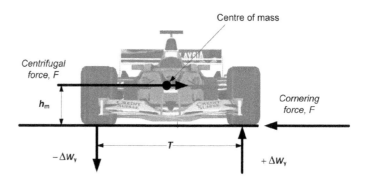

EXAMPLE 1.5

For the same car as in *Examples 1.3* and *1.4* and shown in *Figure 1.17*:

(a) Calculate the cornering force, F, assuming an average coefficient of friction, μ, between the tyre and the road of 1.5.
(b) Determine the maximum total lateral load transfer.
(c) Estimate the velocity that the car can travel around a 100 m radius corner.

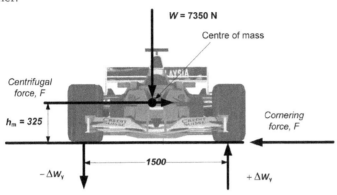

Figure 1.17
Calculating total lateral load transfer during cornering

Weight of car, $W = 7350$ N

(a) From *equation [1.12]*

Maximum cornering force, $F = W \times \mu = 7350 \times 1.5$

$= \mathbf{11\ 025\ N}$

(b) From *equation [1.13]*

Total lateral weight transfer, $\Delta W_y = \pm \dfrac{F h_m}{T} = \pm \dfrac{11\ 025 \times 325}{1500}$

$= \mathbf{\pm 2389\ N}$

(c) From *equation [1.11]*

$$F = \frac{mv^2}{R}$$

Hence $\quad v^2 = \dfrac{FR}{m} = \dfrac{11\ 025 \times 100}{\dfrac{7350}{9.81}}$

$= 1471.5$

$\therefore v = \mathbf{38.4\ m/s\ (138\ km/h)}$

Answer: Cornering force $= \mathbf{11\ 025\ N}$
Total lateral load transfer $= \mathbf{\pm 2389\ N}$
Corner speed $= \mathbf{38.4\ m/s\ (138\ km/h)}$

1.7.1 Cornering and tyre sensitivity

So far, we have referred to **total** lateral load transfer, i.e. the total load that is transferred from the inside wheels to the outside wheels when a car corners. The distribution of this load between the front and rear axles is complex and depends upon the relative stiffness of the front and rear suspensions, suspension geometry, relative track widths, anti-roll bars etc. Altering the proportion of lateral load transfer between the front and rear wheels is an important means by which the suspension is tuned to achieve a balanced car.

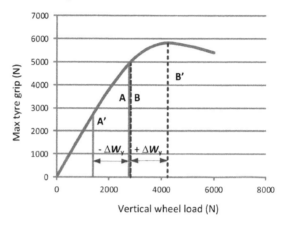

Figure 1.18 Cornering and tyre sensitivity

Consider again the tyre grip curve previously considered in *Figure 1.3* and shown again in *Figure 1.18*:

Lines A and B represent the equal vertical loads on, say, the front wheels during straight-line travel. When the car enters a corner, lateral load transfer, ΔW_y, takes place and this is added to the outer wheel and subtracted from the inner wheel. A' and B' represent the wheel loads during cornering. The significant point to note is that, because of the convex nature of the curve, the sum of the grip at A' and B', after lateral load transfer, is significantly less than that at A and B. In this particular case:

Combined front wheel grip without lateral load transfer = 2 × 4900 N = 9800 N
Combined front wheel grip after lateral load transfer = 2600 + 5800 = 8400 N

Furthermore it can be seen that if the lateral load transfer is increased still further, the grip at A' would continue to reduce significantly and the grip at B' also starts to reduce as the tyre becomes overloaded. We can conclude that, as lateral load transfer increases at either end of a car, the combined grip at that end diminishes. Although total lateral load transfer remains as calculated in *Example 1.5*, the proportion of lateral load transfer at each end of the car can, as already indicated, be engineered to achieve an optimum balance for the car. This topic is dealt with in more detail in *Chapter 5*.

1.8 The g–g diagram

A very useful conceptual tool for visualising the interaction between cornering, braking and acceleration is the **g–g diagram**. It comes in various forms and is alternatively known as the **friction circle** or the **traction circle**. *Figure 1.19* shows a simple form of the g–g diagram for an individual tyre. The diagram indicates the upper boundary to traction in any direction. In this case it indicates that the tyre can support pure acceleration or braking at 1.5g and cornering at 1.4g, but where cornering is combined with braking or acceleration, such as at point A, this figure is reduced. In this case, if the car is accelerating at say 0.75g, it is only able to corner at 1.3g.

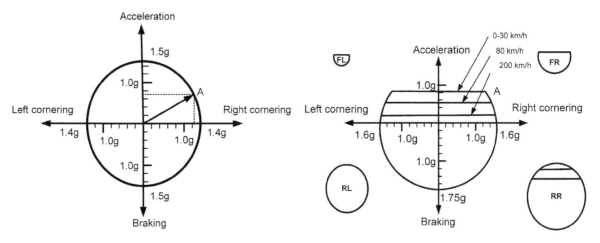

Figure 1.19
The g–g diagram for an individual tyre

Figure 1.20
Traction circle for whole car

As in this example the diagram is not a perfect circle as most tyres will support a little more traction in braking and acceleration than cornering. Also the diameter of the g–g diagram will swell and contract as the individual tyre is subjected to lateral and longitudinal load transfer. A more meaningful form of the g–g diagram is that shown in *Figure 1.20* for the whole car which is obtained by summing the diagrams for the four wheels. The individual wheel loads shown in *Figure 1.20* are for a rear-wheel-drive car which is accelerating out of a left-hand turn, i.e. load transfer to the rear and the right. The whole-car diagram represents the maximum g value that can be achieved in any direction. The flat-top curves in the acceleration zone are the result of rear-wheel-drive and power limitations.

It is the aim of the designer to maximise the size of the traction circle. It is the aim of the driver to keep as close to the perimeter of the circle as possible. *Figure 1.21* shows driver data logged during a race. It can be seen from the above that the driver was braking at up to 1.7g, cornering at 2.2g and accelerating at up to 1.0g.

Figure 1.21
Real traction circle data (produced with ETB Instruments Ltd – DigiTools Software)

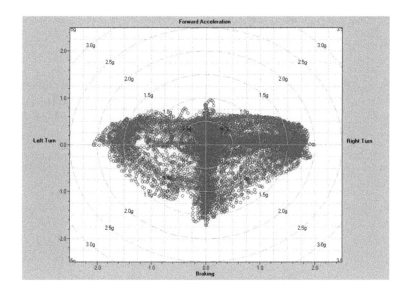

1.9 The effect of aerodynamic downforce

Over the last forty years the performance of racing cars has advanced significantly as evidenced by continually reducing lap times. The biggest single cause is the development of effective aerodynamic packages to produce **downforce**. The objective is to increase traction by increasing the downward force on the contact patch, but without the addition of extra mass. The three major elements that produce downforce are the front wing, the rear wing and the underbody. The design of these elements will be considered in detail in *Chapter 9*. However it is important to appreciate two points:

- Aerodynamic forces are proportional to the square of the velocity of the airflow relative to the car. This means that the designer may need to consider load cases at different speeds as the forces change.
- The penalty paid for downforce generation is increased aerodynamic loss or drag. The designer must decide upon the amount of engine power that can be sacrificed to overcome this drag. This means less power available for accelerating the car and, consequently, a reduced top speed. It follows, therefore, that relatively low-power cars can only run low-downforce aerodynamic set-ups. As the power of the engine increases, a more aggressive downforce package can be adopted.

Table 1.2 shows some examples with approximate figures for guidance:

Table 1.2 Typical downforce classification

Level of downforce	Engine power (bhp)	Max. speed (km/h; mph)	Downforce in *g* values at 180 km/h (110 mph)	Downforce in *g* values at max. speed	Example car
Low	<200	225 (130)	0.5	0.7	Motorcycle-engined single seater
Medium	200–350	250 (150)	0.75	1.4	F3
High	350–700	275 (170)	0.85	2.0	F2
Very high	>700	320 (200)	1.0	3.3	F1

The above two points also lead to different set-ups for different circuits. A typical high-downforce circuit contains lots of fast sweeping corners. A low-downforce circuit consists of tight hair-pin corners joined by fast straights. In such circumstances, even with aggressive wings, little downforce is generated at the corners because the velocity is low and the presence of drag reduces the top-speed on the straights.

FSAE/Formula Student cars present an interesting case in relation to downforce. Average speeds in the range 48 to 57 km/h and maximum speeds of only 105 km/h are at the lower limit of where aerodynamic devices start to become effective. There have been successful teams with wings and successful teams without wings. It is the author's view that well-designed and engineered devices of lightweight construction are, despite a small weight penalty, almost certainly beneficial in the hands of the right driver.

Throughout this book we will occasionally pause and reflect upon the implications of aerodynamic downforce on the design process. We will now consider the effect of aerodynamic downforce on the three elements of racing – acceleration, braking and cornering.

1.9.1 Acceleration and downforce

Downforce has only a relatively small effect on acceleration. For low-power cars the traction-limited stage is relatively short and the car is likely to move into the power-limited stage by the time it has reached about 90 km/h (55 mph). At this speed there is relatively little downforce. Thereafter increased traction provides no benefit for acceleration and the small increase in drag will actually reduce performance. Consequently for low-power low-downforce cars the critical transmission loads will be close to those that occur when accelerating off-the-line.

For high-power high-downforce cars the situation is a little different. Such a car will not reach its power-limit until about 150 km/h (90 mph) by which time it will have developed significant downforce. Transmission loads will increase alongside the growing traction forces. If at 150 km/h the downforce

Race car design

produces say an additional vertical force of 0.7g, the transmission loads will be 70% higher than those off-the-line.

1.9.2 Braking and downforce

Aerodynamic downforce has a significant effect on braking from high speeds. Consider the F1 car in *Table 1.2* – when braking from 320 km/h the car is effectively subjected to 1.0g from gravity plus 3.3g from downforce. Its effective weight therefore becomes its mass × 4.3g. Because of tyre sensitivity the average coefficient of friction, μ, may be reduced from say 1.5 to 1.2. In addition to braking from friction at the tyre contact patch the car will also be slowed by air-braking from aerodynamic drag. As soon as the driver lifts off the throttle, at 320 km/h the car will slow at about 1.5g even without touching the brakes. This braking force is applied at the centre of pressure on the front of the car and not at road level. As this is likely to be close to the centre of mass position it would have little effect on longitudinal weight transfer.

> **EXAMPLE 1.6**
>
> Repeat *Example 1.4* for a F1 car (shown in *Figure 1.22*) braking from 320 km/h assuming:
>
> - aerodynamic downforce of 3.3g which is divided between the wheels in the same proportion as the static loads,
> - drag braking of 1.5g applied at the centre of mass.
>
> (a) Estimate the individual wheel loads during maximum braking assuming an average coefficient of friction, μ, between the tyre and the road of 1.2.
> (b) Calculate the maximum deceleration in both m/s² and g forces.

Figure 1.22
Braking with downforce

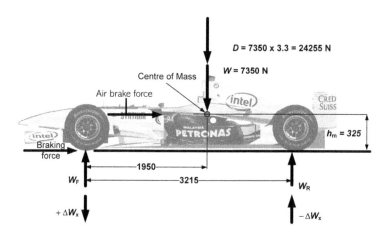

(a) Effective total weight of car, W_T = static weight + downforce

$$= 7350 + 24\,255 = 31\,605 \text{ N}$$

As before: Rear axle load, $W_R = 31\,605 \times \dfrac{1.95}{3.215} = 19\,169 \text{ N}$

Front axle load, $W_F = 31\,605 - 19\,169 = 12\,436 \text{ N}$

From *equation [1.9]*

$$\text{Braking force, } F = W_T \times \mu = 31\,605 \times 1.2$$

$$= \mathbf{37\,926 \text{ N}}$$

From *equation [1.5]*

$$\text{Longitudinal weight transfer, } \Delta W_x = \dfrac{Fh_m}{L} = \pm 37\,926 \times \dfrac{325}{3215}$$

$$= \pm \mathbf{3834 \text{ N}}$$

(b) Front wheel loads, W_{FL} and $W_{FR} = \dfrac{12\,436 + 3834}{2} = \mathbf{8135 \text{ N}}$

Rear wheel loads, W_{RL} and $W_{RR} = \dfrac{19\,169 - 3834}{2} = \mathbf{7668 \text{ N}}$

Air braking force $= 7350 \times 1.5 = 11\,025 \text{ N}$

Total braking force, $F_T = 11\,025 + 37\,926 = 48\,951 \text{ N}$

Mass of car $= \dfrac{7350}{9.81} = 749.2 \text{ kg}$

From *equation [1.4]*

$$\text{Deceleration, } a = \dfrac{F_T}{m} = \dfrac{48\,951}{749.2}$$

$$= \mathbf{65.3 \text{ m/s}^2}$$

$$= \dfrac{65.3}{9.81} = \mathbf{6.7g}$$

Answer: **Wheel loads = 8135 N front and 7668 N rear**
Deceleration = 65.3 m/s² = 6.7g

Comments:
1. Braking at 6.7g is clearly impressive (and stressful for the driver!). However it falls rapidly as the velocity and hence the downforce and drag reduce.
2. The wheel loads calculated above are an important load case for the braking system and the front wheel assemblies, bearings, suspension members etc.

Race car design

1.9.3 Cornering and downforce

Like braking, downforce has a very significant effect on high-speed cornering. The similar large increase in effective weight, and hence grip, combined with a zero increase in mass, mean that the car can sustain high levels of lateral g force. This enables F1 cars to negotiate fast corners such as Eau Rouge at Spa-Francorchamps flat-out at over 300 km/h. Previously we saw that, for a zero-downforce car, the maximum lateral g force on a corner was equal to the average coefficient of friction, μ. It follows that for a car with downforce D:

$$\text{Maximum cornering force, } F = (W+D) \times \mu \quad (N)$$

$$\text{Lateral acceleration in terms of } g \text{ forces} = \frac{(W+D) \times \mu}{W} = (1 + g_{\text{downforce}}) \times \mu$$

From *Table 1.2* (on page 25) for a F1 car at 320 km/h with an average friction, μ = say 1.2, cornering lateral g = (1 + 3.3) × 1.2 = 5.2g.

EXAMPLE 1.7

Repeat *Example 1.4* assuming the car is subjected to aerodynamic downforce, D, of 8100 N also acting at the centre of mass (*Figure 1.23*).

(a) Calculate the cornering force, F, assuming an average coefficient of friction, μ, between the tyre and the road of 1.2. Express this in terms of lateral g forces.
(b) Determine the maximum total lateral load transfer.
(c) Estimate the velocity that the car can travel around a 100 m radius corner.

Figure 1.23
Cornering with downforce

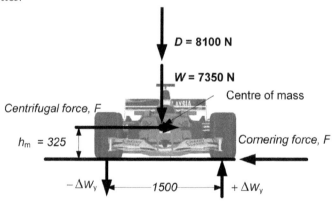

(a) Effective weight of car, W = 7350 + 8100   = 15 450 N

From *equation [1.12]*
Maximum cornering force, $F = W \times \mu$ = 15 450 × 1.2

= 18 540 N

$$\text{In terms of lateral } g \text{ forces} = \frac{18\,540}{7350} = 2.52g$$

(b) From *equation [1.13]*

$$\text{Total lateral weight transfer, } \Delta W_y = \pm \frac{Fh_m}{T} = \pm \frac{15\,450 \times 325}{1500}$$

$$= \pm 3348 \text{ N}$$

(c) From *equation [1.11]*

$$F = \frac{mv^2}{R}$$

$$v^2 = \frac{FR}{m} = \frac{18\,540 \times 100}{\dfrac{7350}{9.81}}$$

$$= 2475$$

$$\therefore v = 49.7 \text{ m/s} \ (179 \text{ km/h})$$

Answer: Cornering force = 18 540 N (2.52g)
Total lateral load transfer = ±3348 N
Corner speed = 49.7 m/s (179 km/h)

Comment:
It can be seen that, compared to the zero-downforce car, the corner speed has increased from 138 km/h to 179 km/h, i.e. a 30% increase.

1.9.4 Effect of downforce on *g–g* diagram

We have seen that downforce has relatively little effect on acceleration but produces significant increases in braking and cornering ability. *Figure 1.24* shows how this may be represented on the *g–g* diagram. It can be seen that, as speed increases, attainable *g* forces in braking and cornering increase. This should be compared to *Figure 1.20* which represents the zero downforce case.

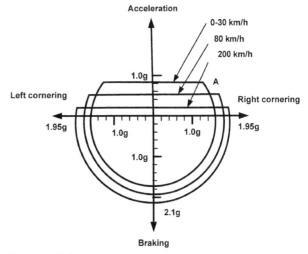

Figure 1.24
A *g–g* diagram for a whole car with downforce

1.10 Racing car design issues

The purpose of this section is to review the preceding topics and to extract those features necessary for a competitive car.

1.10.1 Mass

We have seen that the three elements of racing – acceleration, braking and cornering – all involve either longitudinal or lateral acceleration, and we know from the Newton *equation [1.4]* that in order to maximise these accelerations we need to maximise the force and minimise the mass. The origin of this force, in all cases, is the contact patch between the tyre and the road. We have also seen the phenomenon of **tyre sensitivity** (*Figure 1.3*) which indicates that the effective coefficient of friction between the tyre and the road decreases as the load on the tyre increases. **This means that minimising the mass improves all three elements of racing**. A light car will accelerate, brake and corner better than a heavy one. The Lotus designer, Colin Chapman, is credited with saying:

> 'Adding power makes you faster on the straights. Subtracting weight makes you faster everywhere.'

Provided that a car is adequately stiff, robust and safe it is imperative that the mass is minimised. Where a formula specifies a minimum weight, it is preferable to build the car below that weight and then add strategically placed ballast.

Weight reduction requires discipline in the design and build process and great attention to detail. Stress calculations should be carried out for all key components to optimise the shape and thickness of material. The number and size of bolts should be questioned. Where possible, components should perform more than one function – for example, a **stressed engine** can replace part of the chassis. A bracket or tag can support more than one component. Every gram needs to be fought for and unfortunately this can become expensive as it inevitably leads down the road of expensive materials such as carbon fibre composites.

1.10.2 Position of the centre of mass

The positioning of the wheels and other components needs to ensure that the centre of mass of the car is in the optimum location. Clearly the centre of mass needs to lie as close as possible to the longitudinal centre-line of the car so that the wheels on each side are evenly loaded. This can usually be achieved by offsetting small components such as the battery.

The front-to-rear location of the centre of mass should ensure more weight over the driven wheels to aid traction during acceleration. Thus for a rear-wheel-drive car the centre of mass should be towards the back of the car. A 45:55 or 40:60 front to rear split is generally thought to be about optimum, however this requires wider rear tyres to balance the car during cornering.

With regard to the height of the centre of mass above the ground, we have seen that, because of tyre sensitivity, overall grip reduces as a result of weight transfer during cornering and from *equation [1.13]*:

$$\text{Total lateral weight transfer, } \Delta W_y = \pm \frac{Fh_m}{T}$$

This indicates that the height of the centre of mass, h_m, should be as small as possible to minimise weight transfer. A low centre of mass also produces less roll when cornering which, as we shall see later, means there is less risk of adversely affecting the inclination (camber) of the wheels. **The lowest possible centre of mass is therefore the aim**.

Lastly, items which change their mass during a race, such as the fuel, should be positioned as close to the centre of mass as possible so that the overall balance of the car is not affected as the mass changes.

1.10.3 Engine/drive configuration

Virtually all modern single-seat open-wheel racing cars adopt the rear/mid engine position with rear wheel drive. This facilitates good location of the centre of mass, short (and hence light) transmission of power to the rear wheels and a small frontal area for aerodynamic efficiency. Such cars generally benefit from wider tyres at the rear.

1.10.4 Wheelbase and track

The optimum wheelbase (i.e. length of a car between the centre-lines of the axles) is somewhat difficult to define. In general we know that short-wheelbase cars are nimble and hence good at cornering on twisty circuits, whereas long-wheelbase cars are more stable on fast straights. Hillclimb/sprint cars generally need to negotiate narrower roads with tighter hairpins and hence have evolved relatively short wheelbases (2.0–2.5 m). Circuit racing cars spend more time at higher speeds on wider roads and hence have evolved longer wheelbases (2.5–2.8 m). Modern F1 cars are particularly long (3.1–3.2 m) and the main motivation for this is that the extra length provides a longer floor with which to generate vital downforce. However, all things being equal, a short car is obviously lighter than a long one.

The optimum track (i.e. width of a car between the centres of the wheels) is easier to define. From *equation [1.13]* above we can see that weight transfer reduces as track, T, increases, consequently it usually pays to adopt the widest track that formula regulations allow. A wide track also reduces cornering roll. The regulations are often couched in terms of the maximum overall width of the car which means that the track of the wider rear wheels is a little less than the front.

FSAE/Formula Student presents a special case as far as wheelbase and track are concerned. The narrow and twisting circuits demand a light and highly nimble car. Experience has shown that compact cars perform best, with the wheelbase in the range 1.5–1.7 m and the track around 1.2 m.

> **SUMMARY OF KEY POINTS FROM CHAPTER 1**
> 1. Racing involves optimum acceleration, braking and cornering which all require maximum traction at the contact patch between the tyre and the road.
> 2. Many aspects of racing car design require knowledge of individual vertical wheel loads and the static values are governed by the position of the centre of mass of the car. It is important to be able to calculate this position.
> 3. The wheel loads change as the car accelerates, brakes or corners as a result of load transfer.
> 4. The coefficient of friction between a tyre and the road is not a constant but reduces as the load on the tyre increases. This is known as tyre sensitivity.
> 5. The traction circle is a useful tool to show the interaction between acceleration, braking and cornering.
> 6. The use of aerodynamic downforce is vital for improving track performance and its presence increases wheel loads and hence traction. Its effect is proportional to the square of the velocity of the car.
> 7. We conclude that for optimum performance racing cars should be as light as regulations permit with a low centre of mass and a wide track. A rear/mid engine configuration with rear-wheel drive is favoured.

2 Chassis structure

> **LEARNING OUTCOMES**
>
> At the end of this chapter:
> - You will be able to define the key requirements of a racing car chassis structure
> - You will know the basic types of structure – space-frame, monocoque and stressed skin, and the features of each type
> - You will be able to specify the loads on the chassis structure and understand the need for safety factors
> - You will be aware of analysis techniques for the chassis frame
> - You will know about crash safety structures

2.1 Introduction

The term **chassis** can be used to refer to the entire **rolling chassis**, i.e. including the suspension and wheel assemblies, however this chapter is concerned only with the structural frame of the car. The basic requirements of the structural chassis are:

- to comply with relevant formula regulations,
- to provide secure location for all the components of the car such as the engine, fuel tank, battery etc.,
- to provide sufficient strength and stiffness to resist the forces from the suspension and steering components when the car accelerates, brakes and corners at high g forces,
- to accommodate and protect the driver in the case of a collision and to provide a secure anchorage for the safety harness,
- to support wings and other bodywork when subject to high aerodynamic forces.

The chassis structure is analogous to the human skeleton which keeps all the vital organs in the correct location and provides anchorage for tendons and muscles so that useful movement and work can be carried out.

Over the last fifty years or so there have really been only two major forms of chassis structure:

1. The ***space-frame*** which is a 3-D structure formed from tubes. This is then clad with non-structural bodywork.
2. The ***monocoque*** which consists of plates and shells constructed to form a closed box or cylinder. The monocoque can thus replace some of the bodywork. Modern monocoques are invariably made from carbon fibre composite.

A third form worth mentioning is **stressed-skin construction** which is a hybrid of the first two. This can be an alternative name for a monocoque, but here it is used for a space-frame where some of the members are replaced, or supplemented, by a skin which is structurally fixed to the tubes.

2.2   The importance of torsional stiffness

The torsional deformation of a chassis refers to twisting throughout the length of the car. *Figure 2.1a* shows an underformed space-frame chassis while *Figure 2.1b* shows this chassis subjected to twisting in a torsion test in the laboratory. In practice a chassis will be put in torsion whenever one wheel encounters a high or low spot on the track. It can also occur during cornering as lateral forces cause both horizontal bending and twisting of the chassis, as shown in *Figure 2.1c*.

There are two important reasons why a racing car chassis needs to be torsionally stiff. The first reason concerns the ability to tune the balance of the car effectively, i.e. to achieve a car that has fairly neutral handling without excessive understeer or oversteer. As an example consider a car that suffers from excessive oversteer. This means that, when cornering at the limit, the back-end of the car loses grip first and a spin is the likely outcome. We saw from *section 1.7.1* that, because of tyre sensitivity, as increasing load is transferred across the car, the combined grip at that end of the car is progressively reduced and vice versa. Consequently if we could engineer more load transfer at the front of the car and less at the rear this would reduce front grip and increase rear grip, thus reducing or eliminating the oversteer problem. We can achieve this by stiffening the front suspension in roll and softening the rear suspension in roll. This does, however, mean that the chassis needs to transport lateral cornering loads from, say, the weight of the driver and engine to the front suspension connection points. This causes the chassis to twist. If the chassis has low torsional stiffness it behaves like a spring in series with the front suspension, thus seriously reducing its effective roll stiffness. The ability to tune the balance of the car is thus compromised. It has been shown that for tuning to be at least 80% effective, the torsional stiffness of the chassis needs to be at least approaching the total roll stiffness of the car, i.e. the combined roll stiffness of the front and rear suspensions (*ref. 6*). Milliken and Milliken (*ref. 15*) add:

Chapter 2 **Chassis structure**

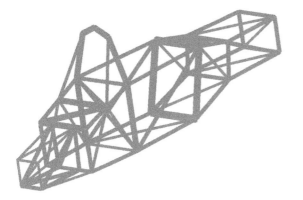

Figure 2.1a
Undeformed chassis

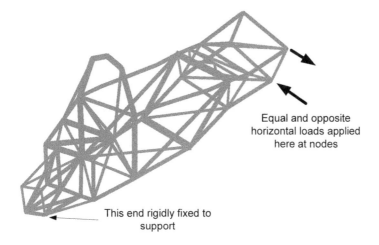

Figure 2.1b
Chassis subjected to torsional twisting

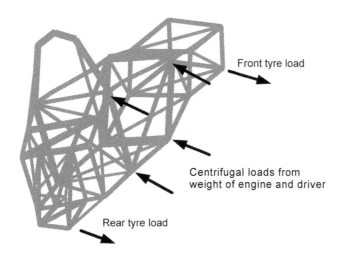

Figure 2.1c
Chassis subjected to torsional twisting and lateral bending from cornering

35

'Predictable handling can best be achieved if the chassis is stiff enough to be safely ignored.'

In practice the total roll stiffness of a racing car increases in proportion to the amount of downforce and hence lateral *g* forces encountered. For a low downforce softly sprung car the total roll stiffness can be as low as 300 Nm/degree, whereas for a F1 car the figure is as high as 25 000 Nm/degree. The required torsional stiffness of the chassis, measured over the length between the front and rear suspension connection points, is thus also in the range 300–25 000 Nm/degree. The higher figures are probably only achievable with carbon fibre monocoque construction. Stiffer is always better, provided not too much weight penalty has been paid, and a 1000 Nm/degree minimum is recommended.

The second reason that torsional stiffness is important is because a flexible chassis stores considerable strain energy. Thus during hard cornering a flexible chassis can 'wind up' like the spring in a mechanical watch. This energy then returns to unsettle the car at the critical point when the driver wants to straighten-up and accelerate out of the corner. (Energy is also stored in the suspension springs, however this is not a problem as it is controlled by dampers.) The absence of chassis damping could result in repeated torsional oscillations.

2.3 The space-frame chassis structure

2.3.1 Principles

The essence of good space-frame design is the use of tubes joined together at **nodes** to form triangles. Each node should have at least three tubes meeting at it. Three such triangles form a **tetrahedron** as shown in *Figure 2.2*.

The centroidal axis of each tube that meets at a node should pass through a single point. All major loads applied to the chassis should be applied only at nodes and ideally also pass through the same single intersection point. The beauty of such a structure is that the tubes are loaded almost entirely in either

Figure 2.2
Tetrahedron with load at node

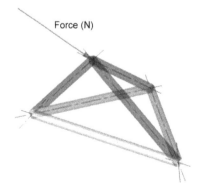

Chapter 2 **Chassis structure**

pure tension or compression, i.e. there is virtually no bending. (The use of 'almost' and 'virtually' in the preceding sentence is due to the fact that, in practice, the nodal joints are generally welded rather than pure pin joints. The members in a truly pin-jointed structure would not contain any bending.) Additional nodes can be created by progressively connecting three more members to existing nodes.

The power of triangulation is clearly illustrated by the 2-D frames shown in *Figure 2.3*.

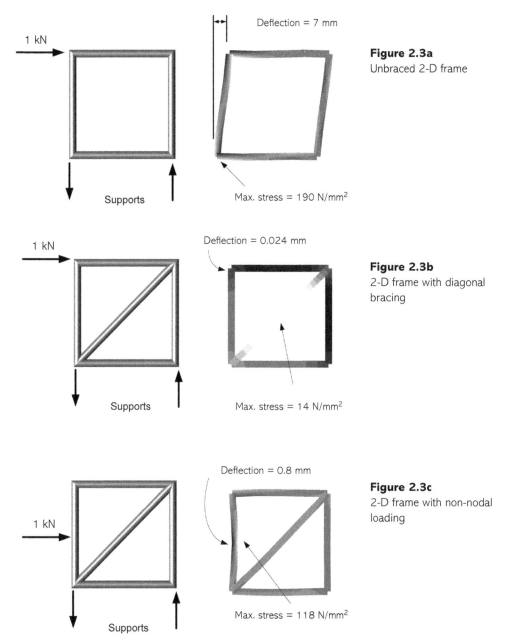

Figure 2.3a
Unbraced 2-D frame

Figure 2.3b
2-D frame with diagonal bracing

Figure 2.3c
2-D frame with non-nodal loading

Figure 2.3a shows an unbraced square 2-D tubular frame supported at the bottom corners and loaded with a 1 kN load at the top corner. The frame is 500 mm square and made from a typical mild steel tube. The left-hand picture shows the undeformed shape and load. The right-hand picture shows the exaggerated deflected shape from a finite element analysis. It indicates the magnitude of the maximum deflection and tensile stress, i.e. 7 mm and 190 N/mm² respectively. *Figure 2.3b* shows the effect of forming triangles by adding a diagonal bracing member. The deflection is reduced to a staggering 0.3% of the unbraced case. Also the maximum tensile stress is reduced to about 7.5%. As the strain energy stored in the structure is proportional to the deflected distance of the load, this is also reduced to 0.3% by introducing the diagonal brace. The members are subjected to almost pure tensile or compressive forces. *Figure 2.3c* shows how, even with a triangulated structure, much of the good work is undone if the load is not applied at a node. Both the maximum deflection and the maximum stress are intermediate between the two former cases.

Although the formation of strong triangulated tetrahedrons is desirable, they are often not ideally suited to the flat-sided shape of most modern racing cars. Another good principle is to aim for prisms of rectangular (or trapezoidal) cross-section where each face of the prism is itself a triangulated rectangle. Such a building-block is shown in *Figure 2.4*. *Figure 2.5* shows several such building blocks with the addition of a roll-bar, and it can be seen that a racing

Figure 2.4
Triangulated prism as a chassis building block

Figure 2.5
Emerging space-frame chassis but some diagonal members had to be removed

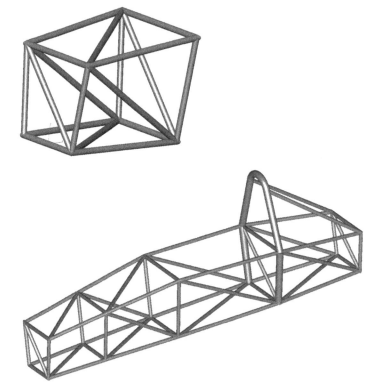

car shape is starting to emerge. However, for practical reasons several key diagonal bracing members have been removed:

- Most significantly, the cockpit opening cannot be diagonally braced.
- Two diagonal members have been removed to permit the driver's legs to extend forwards.
- A diagonal has been removed to allow the engine to be installed and removed.

All of these factors will reduce the torsional stiffness of the chassis and remedial action should be considered. *Figure 2.6a* shows a plan view of a chassis subjected to torsional deformation, where it can clearly be seen that most of the deformation occurs in the cockpit area. Actions that can be considered are:

- increasing the bending stiffness of the upper cockpit side members – particularly in the horizontal plane,
- adding additional higher members between the front and main roll-hoops,
- reducing the effective length of the cockpit side members by adding short diagonals across the corners,
- providing some form of 'picture frame' around the cockpit – possibly in the form of structural side-pods as shown undeformed in *Figure 2.6b*. In this case the torsional stiffness was increased by about 50%.

Figure 2.6a
Deformed chassis under torsion indicating most movement at cockpit opening

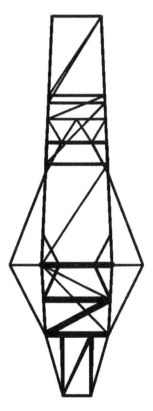

Figure 2.6b
Addition of structural side-pod members increases torsional stiffness by 50%

With regard to the front of the chassis, around the driver's legs, it is possible to introduce one or more bulkheads as shown in *Figure 2.7*. This is particularly desirable if they need to resist non-nodal loads from suspension members. Also many formulae require a substantial front roll-hoop as part of the driver protection structure. Openings to permit the fitting and removal of the engine can be strengthened either by the addition of removable, usually bolted, members and/or by adding bracing to rigid points on the engine itself. Consideration can also be given to making the engine a stressed part of the chassis provided suitable strong connection points exist. In this case the chassis is likely to be split into a front section and a rear section.

Figure 2.7
Typical front roll-hoop bulkhead

2.3.2 Space-frame design process and materials

Because it is important to apply all major chassis loads at (or close to) nodal points it is necessary to know the location of suspension and engine-mounting points before the chassis frame layout can be finalised. It has been said that the process is one of 'joining the dots'. In practice the design of the chassis frame and the suspension tend to proceed in parallel, however the frame designer must be prepared to adjust node positions to suit the suspension.

The initial layout of the chassis frame members must therefore take account of:

1. The particular formula regulations appropriate to the car. These are likely to require minimum ground clearance, size and height of roll-hoops and possibly other driver protection structures. Some formulae such as F1 and FSAE/Formula Student require minimum cockpit sizes by requiring **cockpit profiles** to be inserted into the finished car.
2. The driver dimensions. *Figure 2.8* shows rough dimensions as a starting point to design the cockpit area. The final dimensions used will depend upon the degree to which the driver is reclined, the amount of bend in the knees and the height of the front floor where the pedals are mounted. To maintain an acceptable front line-of-site the front roll-hoop should not be above the height of the driver's mouth.
3. The provision of nodes at (or close to) the points where all high loads attach to the chassis. This includes suspension components, engine mounts and seat belt fixings.

Chapter 2 **Chassis structure**

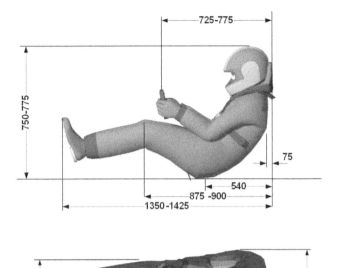

Figure 2.8
Standard driver dimensions for cockpit design

Materials

We have seen that for a well-triangulated space-frame, with loads applied at the nodes, the members are in almost pure tension or compression. Also, as we shall see, many members will be subjected to load reversal under different load cases. Consequently all members need to be designed to carry compressive loads and the only real option for this is the hollow tube. *Table 2.1* lists the main options for space-frame members. Most options are available in several material strength grades and are also delivered in various conditions such as hardened, tempered or annealed. Tubes in the annealed state are easier to bend but usually significantly weaker.

The sizes and properties of standard tubes are given in *Appendix 2*. At the time of writing it was easier to source imperial size tubes than metric. However, note that in *Appendix 2* all the imperial dimensions are given in equivalent metric units.

2.4 Stressed-skin chassis structure

Plate 1 shows the same square frame as *Figure 2.3* but with the diagonal member replaced by a 1 mm thick aluminium sheet. It can be seen that the maximum deflection and stress are of a similar order to the frame with the diagonal member. Clearly the attachment of the plate to the tube requires a high-quality structural bond and this is plainly easier to achieve with square

Race car design

Table 2.1 Options for space-frame tubes

Name	Standard or type	Yield strength (N/mm^2)	Comment
Steel			Density = 7850 kg/m^3 Elastic modulus = 205 000 N/mm^2
Seamless circular tube E355	BS EN 10297-1: 2003	235–470	The UK Motor Sports Association (MSA) requires a minimum yield strength of 350 N/mm^2 for roll cages and bracing. Seamless is also recommended for high stress components such as suspension members.
Welded circular tube	BS EN 10296-1: 2003	175–400	Cheaper than the above and now a consistent and reliable product.
Welded square tube	BS EN 10305-5: 2010	190–420	Square tubes are generally easier to join. Also good for attaching sheets such as floors. More compact than circular tubes and slightly stronger in bending, but about 20% less efficient in compressive buckling.
Alloy steel tube	BS4 T45, AIS1 4130, Osborne GT1000	620–1000	Generally referred to as 'aerospace' tubes, these are made from steel alloys containing elements such as chrome, molybdenum and nickel. They are much more expensive and can require special welding. These will produce the lightest solution where regulations require the chassis frame to resist specific loads.
Aluminium alloy			Density = 2710 kg/m^3 Elastic modulus = 70 000 N/mm^2
6082 – T6	BS EN 754-2: 2008	260	Although this aluminium alloy is about the same strength as mild steel, and is only about a third of the weight, the elastic modulus (stiffness) is also only a third. This reduces its effectiveness in compressive buckling and hence erodes most of the weight advantage. Also welding is more difficult and results in significant strength reduction (50%) in the heat affected zone (HAZ). Ideally significant heat treatment is required after welding.

section tubing. The key is good surface preparation, appropriate curing conditions and the use of a high-quality structural adhesive, often supplemented with rivets or self-tapping screws.

The use of structural flat plates in this way can produce weight saving for such elements as the floor-pan, however, wider use, as an alternative to bodywork, can compromise access, aesthetics and aerodynamics.

2.5 The monocoque chassis structure

2.5.1 Monocoque principles

The monocoque chassis is now universally adopted at professional levels of racing as it has the potential to be lighter, stiffer, stronger and hence safer. It is however more expensive and difficult to build. Whereas the space-frame was based on triangulated tubes in tension and compression, the monocoque is based on plates and shells largely in shear. The triangulated prism building block of *Figure 2.4* is replaced by the plated prism of *Figure 2.9*. Provided it is well restrained at one end, such a structure is very stiff in torsion. As with the

Figure 2.9
Monocoque plated box building block

space-frame, for practical reasons, some faces must be removed or perforated to allow access for the driver and other components. *Plates 2(a)* and *(b)* show the effect on torsional stiffness of removing the end plate. It can be seen that the deflection is increased fourfold, implying that the torsional stiffness is reduced to a quarter. The outer skin of a monocoque is therefore supplemented with internal stiffening ribs and bulkheads. Also thin plate structures are not good for resisting concentrated point loads, so metallic inserts are introduced as reinforcements at key connection points such as where the suspension wishbones attach.

2.5.2 Monocoque materials

Aluminium alloys

In the past, solid aluminium alloy sheet (around 1 mm thick) has been used to form monocoques. The sheet can be bent to form longitudinal boxes or tubes that run the length of the car each side of the driver. These are in turn joined by floor plates, bulkheads and the fire-wall to form a strong overall structure. Sheets can be joined by rivets, welds or adhesive bonding.

A more modern approach is to fabricate the chassis from aluminium honeycomb composite sheet. This consists of two sheets of aluminium bonded to a core of aluminium honeycomb as shown in *Figure 2.10*. Overall thickness is generally about 10–15 mm. This produces a strong and lightweight panel that can be cut, bent and bonded. Clearly it is not possible to produce curves in more than one plane, so aesthetics and aerodynamics may be compromised without the use of separate external bodywork.

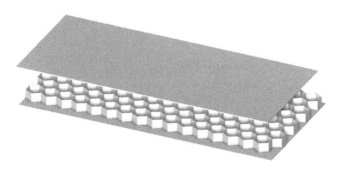

Figure 2.10
Aluminium honeycomb sheet

Carbon fibre composites

Carbon fibre composite is now universally adopted at the professional level as the material of choice for chassis 'tubs' – *Figure 2.11*. Per unit weight, carbon fibre composites are about three times as stiff and strong as structural steel or aluminium. As the word **composite** suggests, it is made up from two distinct materials:

- The **reinforcement** provides most of the strength of the composite and consists of fibres laid in various directions to suit the applied loads. For our purposes the fibres are either **carbon** or **Aramid** (also known as **Kevlar** or **Nomex**). They generally come in the form of a mat of woven or unidirectional fibres. Fibres are available in many grades of varying stiffness and brittleness.
- The **resin matrix** provides the 'body' of the material. It bonds and protects the reinforcement and distributes loads to the fibres. The type of resin determines the capacity of the composite to resist heat. **Phenolic** resins have better fire resistance but **epoxy** resins are stronger and tougher and hence are the obvious choice for a chassis. **Polyester** resins are cheap but lack strength and toughness. **Vinyl ester** resins are intermediate between epoxy and polyester in terms of mechanical strength and cost.
- Carbon fibre composite construction is invariably **anisotropic**, i.e, the fibres are deliberately run in specific directions to produce the most advantageous strength properties.
- Much of the construction consists of a thin skin of carbon fibre composite each side of a lightweight core of Aramid or aluminium honeycomb.

The first step in manufacture is the production of a **pattern** which is a full-sized copy of the chassis in wood or resin. Female moulds or **tools** made of carbon fibre are then cast around the pattern. Chassis tubs are usually made in two halves – an upper and a lower part which are subsequently structurally bonded together. All composite work should be carried out under carefully controlled 'clean room' conditions. Several processes are available for manufacture of the finished parts, two of them being:

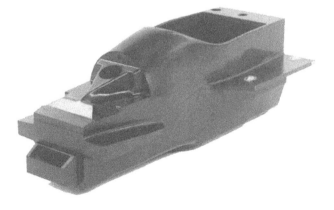

Figure 2.11
Carbon fibre 'tub'. Example shown is a chassis from a Pilbeam MP97 hillclimb car, reproduced with kind permission from Mike Pilbeam

- **Wet lay-up**, which is the simplest approach. Alternate layers of resin and reinforcement are added to the mould and compressed with a hand-roller. The whole is then enclosed in a vacuum bag to compress the composite as the resin cures. The resin is designed to cure at room temperature.
- **Pre-preg** is the professional approach. The reinforcement mats are already impregnated with resin but remain flexible until hot-cured. Different layers are carefully built up before vacuum bagging. The whole is then placed in an **autoclave**, which is a pressurised oven, for curing at a specified temperature, pressure and duration, depending upon the resin used. This process produces the lightest and strongest components but requires expensive equipment.

The production of a high-quality carbon-composite chassis requires considerable skill and experience, to the extent that even Formula 1 teams often use external specialist companies. On the other hand, several FSAE/Formula Student teams have demonstrated considerable success.

2.6 Chassis load cases and safety factors

2.6.1 Load factors

If a chassis is adequately stiff in torsion it is likely to be adequately strong. Nevertheless it is good practice to check the strength of the structure when subjected to peak loads from several load cases. A further complication arises from the fact that a racing car is a highly dynamic object. It is not sufficient to consider the stresses from only the static loads on the car. When cars become airborne, or go over bumps and kerbs, shock loads are transmitted through the suspension, springs and dampers. The actual magnitude of these dynamic loads is very difficult to determine, however the normal design procedure is very simple. It is common practice to apply a dynamic multiplication factor to the static loads. For vertical loads a typical multiplication factor is **3**. This is essentially saying that the mass of the car is subjected to a vertical acceleration of $3g$. For loads that arise from frictional grip, such as those during cornering, or from aerodynamics, a suitable factor is **1.3**. *Table 2.2* lists suggested load cases together with appropriate dynamic multiplication factors.

The factors given in *Table 2.2* are based on *refs 2* and *19*, together with the author's experience. An interesting question arises – is it necessary to check for **load case combinations** such as 'max. vertical load + max. cornering'? In general, structural engineers consider that the likelihood of the same maximum values occurring during such combinations is less likely, and hence tend to use reduced multiplication factors – say 2.0 and 1.1. This means that combined cases are not usually critical.

Table 2.2 Load cases and dynamic factors

Load case	Dynamic multiplication factor
Max. vertical load	3.0
Max. torsion (diagonally opposite wheels on high-spots)	1.3 on vertical loads
Max. cornering	1.3 on vertical and lateral loads
Max. braking	1.3 on vertical and longitudinal loads
Max. acceleration	1.3 on vertical and longitudinal loads

2.6.2 Material factors

In addition to the factors on loads given in *Table 2.2*, it is necessary to apply a safety factor to the material strength. A value of **1.5** is suggested. This takes account of factors such as:

- quality of the material,
- small errors in component dimensions,
- defects in the component such as lack of straightness,
- loads not being applied on the centroid of the component.

Some designers may use a slightly higher safety factor for *safety critical* and *mission critical* components and a lower factor for other components – say 1.6 and 1.4 respectively. An example of a component in the former category is a suspension wishbone member and in the latter category is a diagonal bracing member whose primary function is to increase torsional stiffness. It may also be appropriate to vary the safety factor for different materials. Thus steel and aluminium obtained from a reliable source may attract lower safety factors than say home-made fibre composites.

2.7 Design of structural elements

2.7.1 Components in tension

We have seen that triangulated structures with nodal loads contain members that are largely loaded in pure tension or compression. For components in tension the above material safety factor (say 1.5) is simply applied to the material **yield stress**, σ_y (or in the case of aluminium, which does not have a clearly defined yield point, to the **0.2% proof stress**). Hence:

$$\text{Tensile stress}, \sigma_t = \frac{\text{Force}, F_t}{\text{Area}, A} \leq \frac{\text{Yield stress}, \sigma_y}{1.5}$$

$$\therefore \text{Min. area}, A = \frac{1.5 \times F_t}{\sigma_y} \quad [2.1]$$

2.7.2 Components in compression

Slender **compression members**, or struts, fail in buckling often long before the yield stress is reached. A reasonable indication of the strength of such members is given by the **Euler buckling load**. For pin-ended components this is:

$$\text{Euler buckling load, } P_E = \frac{\pi^2 EI}{L^2}$$

where
E = Modulus of elasticity
I = Second moment of area
L = Effective length.

In this case the material safety factor (say 1.5) is applied to the Euler formula:

$$\text{Allowable Euler buckling load} = \frac{P_E}{1.5} = \frac{\pi^2 EI}{1.5 L^2} \qquad [2.2]$$

For pin-ended struts the effective length is the distance between nodes. In a frame with fully welded joints the effective length of compression members can be taken as 0.85 × distance between nodes.

An application of this formula is demonstrated in *Example 2.1* on page 49.

2.7.3 Components in bending

As in *Figure 2.3c*, if a force is applied to a member some distance away from a triangulated node it will cause a **bending moment** in that member which, in turn, generates a bending stress. Again for a material safety factor of 1.5 we get:

$$\text{Bending stress, } \sigma_b = \frac{\text{bending moment, } M}{\text{elastic modulus, } Z} \leq \frac{\text{yield stress, } \sigma_y}{1.5}$$

$$\therefore \text{Min. elastic modulus} = \frac{1.5 \times M}{\sigma_y} \qquad [2.3]$$

The elastic modulus is a geometric property based on the cross-section of the member. Values are given in tables for standard tubes.

If a structural member is subjected to both a bending moment *and* a tension force at the same time, we can combine *equations [2.1]* and *[2.3]* to get:

$$\text{Max. stress, } \sigma_b = \frac{F_t}{A} \pm \frac{M}{Z} \leq \frac{\text{Yield stress, } \sigma_y}{1.5} \qquad [2.4]$$

For more information on the calculation of bending moments and the design of structural elements see *ref. 22*.

Race car design

2.8 Chassis stress analysis

2.8.1 Hand analysis

Simple hand analysis techniques can be useful for checking the strength of principal members in a space-frame chassis. Consider the chassis shown in *Figure 2.12a*. The aim is to determine the forces in the main members **a**, **b** and **c** when the chassis is subjected to the maximum vertical load case.

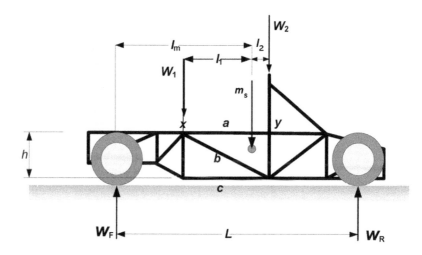

Figure 2.12a
Hand analysis of space-frame chassis

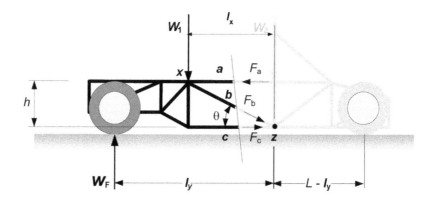

Figure 2.12b
Application of the method of sections

The first step is to determine the weight of the car which is applied at the centre of mass, taking account of the dynamic multiplication factor. It is appropriate to consider the **sprung mass**, m_s, in this case, i.e. the mass of the wheel assemblies with half the wishbones excluded as these are carried directly through the tyres to the ground without passing through the chassis frame. The front and rear wheel loads are then determined. For symmetrical vertical loading, half the car design weight is then apportioned to the nearest adjacent frame nodes at one side of the car. The load at each node, W_1 and W_2, is inversely proportional to its distance from the centre of mass. This is a conservative approach as in reality the sprung mass of the car is distributed more widely throughout the frame. For a dynamic multiplication factor of 3:

$$\text{Design weight of car, } W = m_s \times 3 \times 9.81 \text{ N}$$

Moments about W_F Rear wheel load, $W_R = \dfrac{W}{2} \times \dfrac{l_m}{L}$

From vertical equilibrium Front wheel load, $W_F = \dfrac{W}{2} - W_R$

Load at **x**, $W_1 = \dfrac{\dfrac{W}{2} \times l_2}{(l_1 + l_2)}$

Sum vertical forces Load at **y**, $W_2 = \dfrac{W}{2} - W_1$

The next step is to use a technique known as the **method of sections** to determine the forces in **a**, **b** and **c**. The members in question are assumed to be cut by an imaginary line as shown in *Figure 2.12b*. The structure to the right of this line is ignored. All forces to the left of the line must be in equilibrium with the forces shown as arrows on the cut members **a**, **b** and **c**. All three of these forces are unknown at this stage but if we take moments about a point where two of them intersect, i.e. z, these two are eliminated leaving the force in **a** as the only unknown.

Moments about z $W_F \times l_y = (W_1 \times l_x) + (F_a \times h)$ Hence F_a

Moments about x $W_F \times (l_y - l_x) = F_c \times h$ Hence F_c

If members **a** and **c** are horizontal then the vertical component of the force in **b** must provide vertical equilibrium.

Sum vertical forces $W_F - W_1 = F_b \times \sin\theta$ Hence F_b

---

EXAMPLE 2.1

Figure 2.13a shows a space-frame racing car chassis. It has a fully laden sprung mass of 530 kg.

1. Determine the forces in members **a**, **b** and **c**.
2. Check the suitability of using 25 mm outside diameter circular tubes with a wall thickness of 1.5 mm, given:

- Cross-sectional area = 110.7 mm^2

- Second moment of area = 7676 mm^4
- Yield stress = 275 N/mm^2
- Elastic modulus = 200 000 N/mm^2.

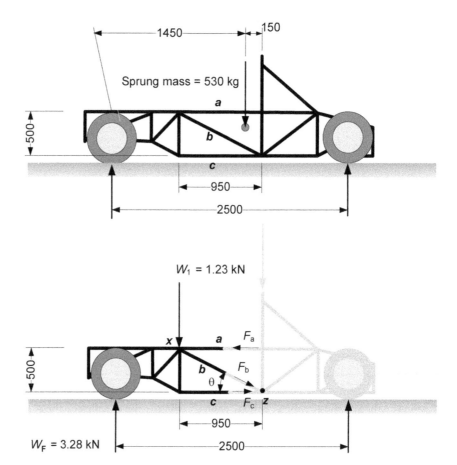

Figure 2.13a Analysis of space-frame chassis

Figure 2.13b Applying the method of sections

Solution

1. Design weight of car, $W = 530 \times 3 \times 9.81/10^3$ N $= 15.6$ kN

Moments about W_F

$$\text{Rear wheel load, } W_R = \frac{W}{2} \times \frac{l_m}{L} = \frac{15.6}{2} \times \frac{1450}{2500}$$

$$= 4.52 \text{ kN}$$

From vertical equilibrium

$$\text{Front wheel load, } W_F = \frac{W}{2} - W_R = \frac{15.6}{2} - 4.52$$

$$= 3.28 \text{ kN}$$

$$\text{Load at } \pmb{x}, W_1 = \frac{\frac{W}{2} \times l_2}{(l_1 + l_2)} = \frac{\frac{15.6}{2} \times 150}{(800 + 150)}$$

$$= 1.23 \text{ kN}$$

Sum vertical forces

$$\text{Load at } \pmb{y}, W_2 = \frac{W}{2} - W_1 = \frac{15.2}{2} - 1.23$$

$$= 6.37 \text{ kN}$$

Method of sections (Figure 2.13b)

Moments about z

$$W_F \times l_y = (W_1 \times l_x) + (F_a \times h)$$
$$3.28 \times 1600 = (1.23 \times 950) + (F_a \times 500)$$
$$F_a = 8.16 \text{ kN compression}$$

Moments about x

$$W_F \times (l_y - l_x) = F_c \times h$$
$$3.28 \times (1600 - 950) = F_c \times 0.5$$
$$F_c = 4.26 \text{ kN tension}$$

$$\theta = \tan^{-1}\frac{500}{950} = 27.8°$$

Sum vertical forces

$$W_F - W_1 = F_b \times \sin\theta$$
$$3.28 - 1.23 = F_b \times \sin 27.8°$$
$$F_b = 4.40 \text{ kN tension}$$

2. Check strength of members

Member $\pmb{a}$ $\quad F_a = 8.16 \text{ kN compression}$

Equation [2.2]

$$\text{Euler buckling load, } P_E = \frac{\pi^2 EI}{1.5 l^2}$$

where

E = Elastic modulus = 200 000 N/mm^2
I = Second moment of area = 7676 mm^4

Welded joints

l = Effective length = 0.85 × 950
$\quad$ = 807.5

1.5 = Material safety factor

$$P_E = \frac{\pi^2 \times 200\,000 \times 7676}{1.5 \times 807.5^2} = 15\,491 \text{ N}$$

| | | = 15.5 kN > 8.16 kN   OK |

*Member **b*** 　　　　　　F_b = 4.40 kN tension

$$\text{Tensile stress, } \sigma_t = \frac{\text{force}}{\text{area}} = \frac{4400}{110.7}$$

$$= 39.7 \text{ N/mm}^2$$

$$\text{Allowable stress} = \frac{\text{yield stress}}{\text{material safety factor}} = \frac{275}{1.5}$$

$$= 183 \text{ N/mm}^2 > 39.7 \quad \text{OK}$$

*Member **c*** 　　　　　　F_c = 4.26 kN tension

By inspection of above   **Stress OK**

Comment:

From this load case, consideration could be given to reducing the size of the members, particularly with respect to the tension members.

2.8.2 Computer analysis

Stress analysis

Chassis frame stress analysis is generally performed with a proprietary **finite element** computer package. The first use of such programs was in the aerospace industry and many variants are available. These range from complex dedicated FEA packages, capable of handling non-linear materials and crash analysis, to simpler add-ons to 3-D modelling packages. Most are expensive but useful information, particularly for space-frames, can be obtained from freely available software such as LISA (*ref. 12*) The following illustrates the use of LISA to analyse the same simple space-frame used in *Example 2.1*.

Figure 2.14 gives a specimen screen from LISA showing the chassis frame model. The following points are of note:

1. The first step is to input the nodal coordinate points.
2. The material properties and sizes are defined for each different tube.
3. The nodes are then connected by **elements**. In this case they are line elements and each one has particular material properties associated with it.

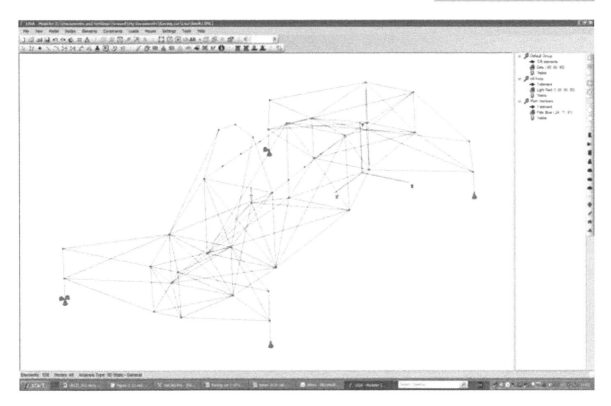

Figure 2.14
Specimen screen of chassis model from LISA finite element program

4. In most cases the elements are assumed to be rigidly connected (welded) at the ends, but the suspension wishbones are defined as pin-ended truss members.
5. The suspension **pushrods** and springs that connect the wheel assemblies to the chassis present an interesting challenge. It is possible to build relatively elaborate models of rotating **bell-cranks** etc.; however in this case a much simpler approach has been adopted. The elastic modulus value of the pushrods has been significantly reduced so that they behave as springs. They have been connected to the node on the chassis where the bell-crank pivots occur.
6. An additional node has been created at the location of the centre of mass and this has been connected to adjacent nodes, again with members of low elastic modulus so that their presence will have little effect on the loads in the main chassis.
7. Support restraints require careful consideration. It can be seen that the wheel uprights have been extended down to ground level to form contact patches. The car should be provided with the minimum number of restraints required to provide stability. In this case each 'wheel' has been restrained vertically. Both contact patches on the outside of the corner have been restrained laterally and one rear contact patch restrained longitudinally. The addition of any further unnecessary restraints is likely to cause arching across the car, which will affect the forces in the tubes.

8. Factored design loads are now added at the centre-of-mass node. In addition to a vertical component to simulate the sprung mass, a lateral component can simulate cornering and longitudinal components can simulate acceleration and braking.
9. For the maximum torsion case, a vertical load is added and diagonally opposite wheel contact patches are displaced upwards until the reaction at one of the other wheels is zero.

Results

After solving, LISA produces a series of contoured deformed images together with a file of tabulated output giving nodal displacements and relevant loads and stresses for each element. *Figures 2.15a–c* show results for three load cases with the deformations exaggerated. The forces in the three main side members that were analysed in *Example 2.1* are also indicated. It can be seen that, for the maximum vertical load case, the values are similar to those obtained by hand calculation. The differences are probably explained by the fact that the load was applied in a slightly different way.

Simple finite element tools often do not check compressive buckling. It is not appropriate simply to check the maximum compressive stress against the yield stress of the material. The compressive load should be compared with the Euler buckling load as in *Example 2.1*.

Stiffness analysis

It has been pointed out that high torsional stiffness is an important objective for a racing car chassis and a computer finite element analysis provides a good method of obtaining the torsional stiffness to compare with the total roll stiffness of the suspension. The procedure is to fix one end of the frame at say the nodes where the rear suspension bell-cranks attach and apply a torque at the other end where the front suspension attaches. The torque applied is an arbitrary value and is not related to actual loads on the car. The reason for this is that we want to know how the torsional stiffness of the chassis compares to other cars.

EXAMPLE 2.2

Plate 3 shows the model and the output from such a finite element analysis. Calculate the torsional stiffness of the chassis in Nm/degree.

$$\text{Torque applied} = 1000 \times 0.270$$

$$= 270 \text{ Nm}$$

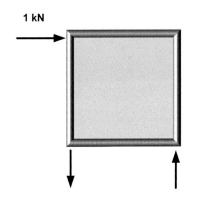 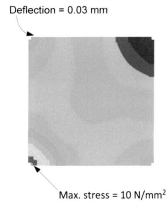

Deflection = 0.03 mm

Max. stress = 10 N/mm²

1 kN

Plate 1
2-D frame with stressed aluminium plate

(a)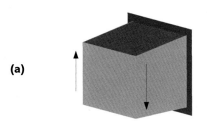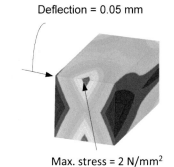

Deflection = 0.05 mm

Max. stress = 2 N/mm²

Plate 2
(a) Torsion on plated prism;
(b) torsion with open end

(b)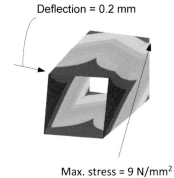

Deflection = 0.2 mm

Max. stress = 9 N/mm²

Plate 3
Torsional stiffness test

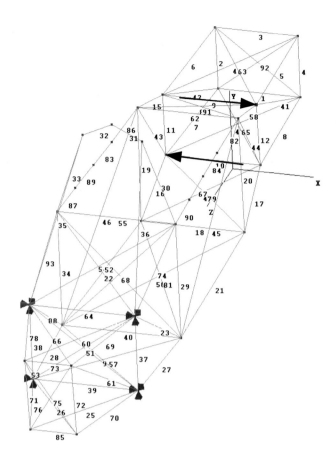

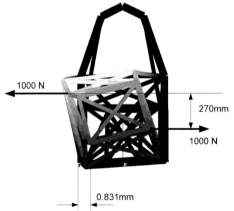

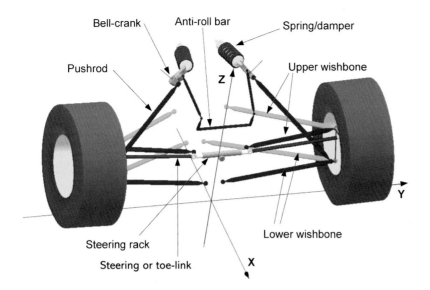

Plate 4
Elements of front suspension

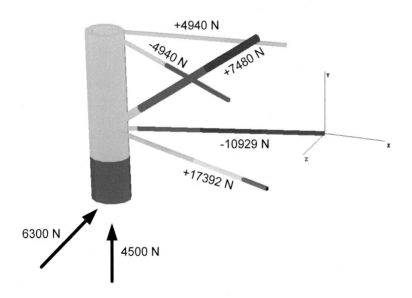

Plate 5
Wishbone forces from computer analysis

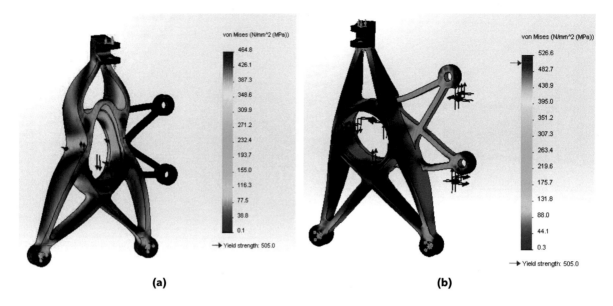

Plate 6
Finite element analysis of front upright

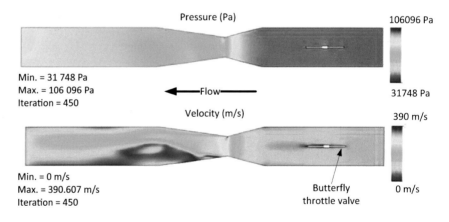

Plate 7
CFD investigation of Venturi restrictor

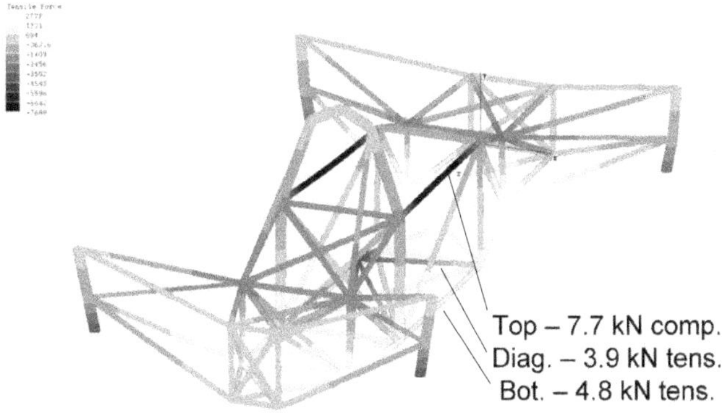

Chapter 2 **Chassis structure**

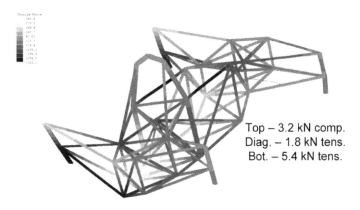

Figure 2.15a
Maximum vertical load case

Top – 7.7 kN comp.
Diag. – 3.9 kN tens.
Bot. – 4.8 kN tens.

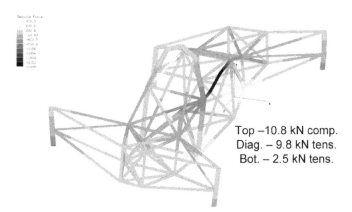

Figure 2.15b
Maximum cornering load case

Top – 3.2 kN comp.
Diag. – 1.8 kN tens.
Bot. – 5.4 kN tens.

Figure 2.15c
Maximum torsion load case

Top –10.8 kN comp.
Diag. – 9.8 kN tens.
Bot. – 2.5 kN tens.

$$\text{Angular rotation} = \tan^{-1} \frac{0.831}{270}$$

$$= 0.176°$$

$$\text{Torsional stiffness} = \frac{270}{0.176}$$

$$= \mathbf{1534 \text{ Nm/degree}}$$

Comments:

1. *Section 2.2 suggested that a target value for torsional stiffness should be the total roll stiffness of the car with a suggested minimum of 1000 Nm/degree. The above figure therefore looks acceptable.*
2. *The above calculation shows the total twist angle (0.176°) throughout the length of the chassis. It can be instructive to plot the growth in twist angle at several intermediate increments to see if there are specific sections of the chassis that are particularly flexible in torsion. In this case the cockpit opening is responsible for most of the twist.*
3. *It is often worth repeating the analysis with the size of certain members changed and possibly other members added or removed in order to be clear about their impact on torsional stiffness. A possible target is to maximise the torsional stiffness:mass ratio. Additional bracing around the cockpit opening could be investigated in this way.*

2.9 Crash protection

The principle behind driver protection in the event of a collision is the same as that employed in road cars, i.e. enclose the driver in a strong safety cell and surround the cell with crumple zones of energy-dissipating material. Most formula regulations either contain rules about the size of members in the safety cell, such as the roll-hoop, or specify specific loads that need to be resisted. Designs must generally be supported by approved calculations and physical tests.

The regulations for frontal (and rear) impact-absorbing structures generally require the car to be stopped from a specified speed without subjecting the driver to excessive *g* force. This requires an **impact attenuator** with particular qualities. It must start to crumple under a well-defined force and then continue to resist this force as it yields over a large displacement. This ensures that the area under the load/displacement graph, which is a measure of energy dissipation, is maximised. Kinetic energy is converted into heat. Various specialised foams are available for this purpose as well as aluminium

Figure 2.16
Aluminium impact attenuator before and after crush testing

honeycomb. *Figure 2.16* shows an aluminium honeycomb attenuator before and after crush testing. *Figure 2.17* shows a graph of test data which indicates a high initial peak stress, followed by a long zone of yielding at almost constant load. The initial peak is a result of the load first causing buckling of the honeycomb material. Most formula regulations permit such peaks for very short periods, however the peak can be removed by pre-crushing the material by a few millimetres before installation. The energy absorbed by the attenuator is the area under the force/displacement curve.

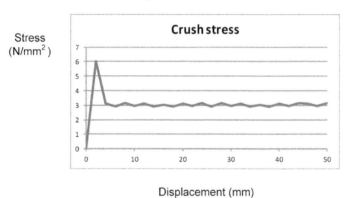

Figure 2.17
Typical crush stress curve for aluminium honeycomb

The design procedure is as follows, given:

Total mass of car = m
Initial velocity of car = u
Maximum number of g forces = G

The first step is evaluate the distance over which the car must be slowed in order to stay within the g force limit and hence to determine the required original length of the attenuator.

$$\text{Deceleration rate, } a = -9.81 \times G$$

From standard equation of motion
$$v^2 = u^2 + 2as$$
where v = final velocity = 0
u = initial velocity
s = distance

Race car design

Rearranging
$$s = \frac{u^2}{2a}$$

From manufacturer's data
Effective length of attenuator = 70% of original length

$$\text{original length} = \frac{s}{0.7}$$

We now need to specify the type of honeycomb and its frontal area so that the attenuator provides the correct force whilst crushing. Manufacturer's data for a range of types is as shown in *Table 2.3 – ref. 10*.

Table 2.3 'HexWeb' honeycomb crush strength (www.hexcel.com)

Type	Density (kg/m³)	Crush strength (N/mm²)
1/4-5052-3.4	54	1.03
1/4-5052-4.3	69	1.59
1/4-5052-5.2	83	2.31
1/4-5052-6.0	96	2.97
1/4-5052-7.9	127	5.00

The crush strengths given in *Table 2.3* are static values and they increase with the speed of impact as shown approximately in *Figure 2.18*.

Figure 2.18
Increase in crush strength with velocity

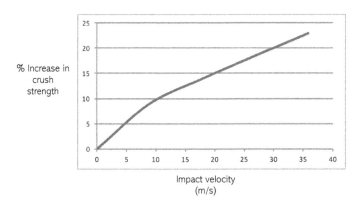

From Newton   Force = Ma

Also   Force = crush strength × frontal area

Rearranging   Frontal area, $A = \dfrac{M \times a}{\text{crush strength}}$

EXAMPLE 2.3

FIA Formula 3 regulations state that a 650 kg car must be stopped from 12 m/s without exposing the driver to a deceleration of more than 25g.

Determine the dimensions of a suitable aluminium honeycomb impact attenuator.

Solution

$$\text{Deceleration rate, } a = 9.81 \times G = 9.81 \times 25$$
$$= 245.3 \text{ m/s}^2$$

$$\text{Distance, } s = -\frac{u^2}{2a} = \frac{12^2}{2 \times 245.3}$$
$$= 0.294 \text{ m}$$

$$\text{Original length of attenuator} = \frac{s}{0.7} = \frac{0.294}{0.7}$$
$$= 0.420 \text{ m}$$

$$\text{Force, } F = ma = 650 \times 245.3$$
$$= 159\,445 \text{ N}$$

Try type 1/4-5052-6.0 from Table 2.3

$$\text{Crushing strength at 12 m/s} = 2.97 \times 1.11$$
$$= 3.3 \text{ N/mm}^2$$

$$\text{Frontal area, } A = \frac{159\,445}{3.3} = 48\,320 \text{ mm}^2$$

$$\text{For a width of 300 mm, height} = \frac{48\,320}{300} = 161 \text{ mm}$$

Check

$$\text{Kinetic energy of moving car} = 0.5\,mu^2 = 0.5 \times 650 \times 12^2$$
$$= 46\,800 \text{ Joules}$$

$$\text{Energy absorbed by attenuator} = F \times s = 159\,445 \times 0.294$$
$$= 46\,880 \text{ Joules} \quad \textbf{check}$$

Answer:
Use an aluminium honeycomb *type 1/4-5052-6.0* attenuator 420 mm long × 300 mm wide × 161 mm high

SUMMARY OF KEY POINTS FROM CHAPTER 2

1. The chassis frame must support the static and dynamic loads from the suspension and other components as well as protecting the driver in the event of a collision.
2. Chassis frames should be torsionally stiff if a car is to handle well.
3. The two main forms of chassis are tubular space-frames and monocoque tubs.
4. Space-frames should be fully triangulated where possible and all external loads applied at, or close to, nodes. Strengthening may be required where this is not possible.
5. Two commonly used materials for monocoque construction are aluminium honeycomb sheet and carbon fibre composite.
6. Static loads must be increased to allow for dynamic effects and a safety factor should also be applied to material strength.
7. Useful information on the loads in principal members can be obtained from simple hand calculations but for a full analysis the modern approach is to use a finite element package.
8. For safety reasons, the driver should be protected by a strong safety cell which is itself surrounded by energy-dissipating material.

3 Suspension links

> **LEARNING OUTCOMES**
>
> At the end of this chapter:
> - You will know what is required from a racing car suspension in order to optimise performance
> - You will know how to design a double wishbone suspension and understand how geometry changes affect the ability to control wheel camber and other important characteristics
> - You will learn how suspension geometry can be used to control the pitching of a car during acceleration and braking – so called anti-squat and anti-dive
> - You will be able to calculate loads in the suspension and select suitable structural members
> - You will be aware of why particular wishbone geometries have been adopted in a range of different cars

3.1 Introduction to racing car suspensions

A well-designed suspension is vital for good performance of a racing car. Tony Pashley (*ref. 18*) defines the challenge well:

> *'I should point out here that our purpose in designing the suspension and steering is to attempt to maintain the wheels at the optimum angle to the road surface at all times and under all conditions … . This is the Holy Grail of the designer and is, in practice, unattainable.'*

Plate 4 shows a typical racing car front suspension configuration and indicates some of the terms used. This is what is known as a **double-wishbone** suspension and it has become universally adopted throughout racing. Consequently, although many other forms of suspension mechanism exist, this book will focus only on the design of the double-wishbone and its variants. The rear suspension can be almost identical but with a fixed toe-link as, of course, there is no need to provide steering. The double wishbone is a form of **independent suspension**, which, unlike a solid beam axle, allows each individual wheel to react independently to local irregularities in the road surface. This chapter is concerned only with the design of the wishbones. *Chapter 4* will look at springs, dampers, bell-cranks and the ***anti-roll bar***. *Chapter 6* is concerned with the assembly inside the wheel and the steering components.

The upper and lower wishbones form the mechanism that joins the wheels to the chassis structure. With the toe-link in place, but the spring/damper system disconnected, it is possible to move the wheel freely along a pre-determined path in relation to the chassis frame. The nature of this movement path is very important to the handling of the car and it is sensitive to the location of the connection points on the chassis. Small changes to the geometry of the links can make important changes to the movement (or **kinematics**) of the wheel. The nature of this movement can be determined by calculation, drawing or a scale model; however the modern approach is to use a suspension design computer package. *Plate 4* and many subsequent diagrams in this chapter are from a package called **SusProg3D Designer** (*ref. 27*) and several others exist.

At each axle there are two basic relative movements between the chassis and the wheels:

1. **Bump and rebound** – in which the two wheels move in the same direction.
2. **Roll** – in which the wheels move in opposite directions.

These are illustrated in *Figures 3.1a–d*. All other movements can be considered to be combinations of these two basic types. If you look carefully at *Figure 3.1* it is clear that the vertical inclination, or **camber**, of the wheels is affected and this can have an important influence on grip.

3.2 Wheel camber and grip

Wheel camber angle, ϕ, is defined as the angle between the plane of the wheel and the vertical. Camber is said to be **positive** if the top of the wheel is leaning outwards relative to the vehicle as shown in *Figure 3.1c* and **negative** if leaning inwards as shown in *Figure 3.1b*. Thus in the case of roll – *Figure 3.1d* – the heavily loaded outer wheel has positive camber and the lightly loaded inner wheel has negative camber.

Figure 3.2 gives typical values derived from physical tyre tests and shows how both lateral and longitudinal grip are influenced by camber angle. It can be seen that, in this case, peak grip for the heavily loaded outside wheel in cornering occurs at about –1°, and that it falls off quickly when the wheel goes into positive camber. However for the lightly loaded inside wheel the peak grip occurs under positive camber. This arises because of the definition of camber angle. In fact, for maximum grip, in both cases the top of the wheel should lean towards the turn centre. In *Figure 3.1d* both the inner and outer wheel are *wrongly* cambered for peak grip. For acceleration and braking, as expected, the peak grip occurs at zero camber when the size of the contact patch is maximised.

Chapter 3 **Suspension links**

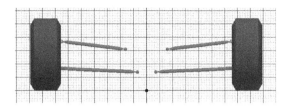

Figure 3.1a
Static

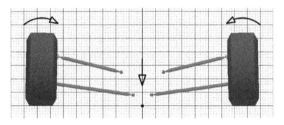

Figure 3.1b
Bump

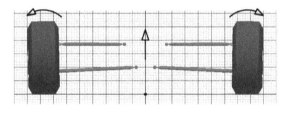

Figure 3.1c
Rebound

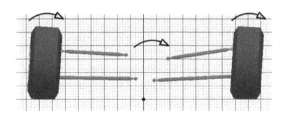

Figure 3.1d
Roll

3.3 The double wishbone suspension – front view

3.3.1 Parallel equal-length double wishbones

The starting point for understanding the double wishbone suspension is to consider what happens under the two main movements of bump and roll when the wishbones are both parallel and of equal length. *Figure 3.3* shows such a case together with the results from 50 mm bump and 4° roll. It can be seen that, as expected for a parallel linkage, 4.0° of roll produces 4.0° of

Race car design

Figure 3.2
The effect of camber on wheel grip

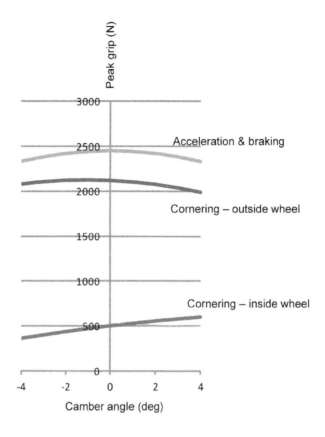

Figure 3.3a
Parallel equal-length wishbones in roll

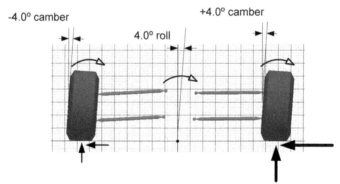

Figure 3.3b
Parallel equal-length wishbones in bump/rebound

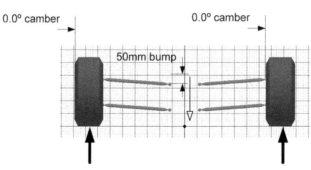

adverse camber change on both wheels. However, under bump (and rebound) there is no camber change.

There are several options open to the designer to ameliorate this problem under roll:

- Stiffen the springs and anti-roll bars to reduce the amount of roll and hence adverse camber change. This may be a viable option in part however, as we shall see later that springs should be as soft as possible to maximise grip.
- Add static wheel camber. If, in this case, the wheels had started out with −4.0° of static camber, then the important outer wheel would end up at zero camber after rolling. Unfortunately the inner wheel would end up at +8.0° and tyre wear may be a problem.
- Use converging instead of parallel wishbones.
- Use a shorter top wishbone instead of one of equal length.

These last two options will now be explained in more detail.

3.3.2 Converging equal-length double wishbones

Figure 3.4a shows a single wheel with non-parallel converging wishbones. Start by considering the chassis to be fixed and the wheel rising as though it is passing over a bump. It is clear that the top wishbone will pivot about point **a** and consequently that point **b**, on the wheel, will move perpendicular to the wishbone as shown by the arrow. Likewise point **d** will move perpendicular to the bottom wishbone which pivots about **c**. This means that the movement of the wheel, relative to the chassis, is effectively a circular path with its centre at the projected intersection point of the wishbones. This point is known as the **instant centre**. The word 'instant' is used because, as the geometry of the links change, the point will move.

The above implies that for small movements the suspension can be considered to be a **swing arm** rigidly connected to the wheel and pivoted about the instant centre as shown in *Figure 3.4b*. The full terminology for this is **front view swing arm** or **fvsa** and its length is L_{fvsa}. In reality, however, when the car rolls in cornering, the tyre stays in contact with the road and rotation can be considered to be about the centre of the tyre contact patch. Therefore if a line is drawn from the centre of the contact patch to the instant centre, an imaginary point on the chassis, at the instant centre, will move perpendicular to this line as shown by the arrow in *Figure 3.4b*.

Figure 3.4c shows both wheels and a rectangle to represent the chassis. It is clear that the chassis will rotate about the intersection point between the two lines joining the respective contact patches and instant centres. This point is very important and is known as the **roll centre**. Initially, for a symmetrical suspension set-up, the roll centre will lie on the centre-line of the car but, as the instant centres move, it will migrate to a new location. The magnitude of this movement depends upon the geometry of the wishbone links, particu-

Figure 3.4a
Converging wishbones and instant centre

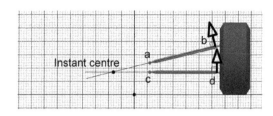

Figure 3.4b
Converging wishbones and swing arm length

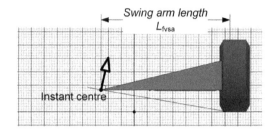

Figure 3.4c
The roll centre

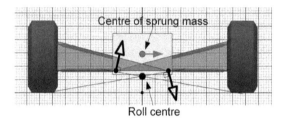

larly the location of the chassis connection points. Excessive movement of the roll centre during cornering, as we shall see in *Chapter 5*, adversely affects the handling balance of the car and hence should be avoided. A maximum movement of about 100 mm laterally and 50 mm vertically is thought to be acceptable but many designers will aim for much less, particularly vertically.

The relationship between the roll centre and the centre of mass of the sprung chassis is fundamental. As shown in *Figure 3.4c* the centripetal force from the sprung mass acts horizontally through the centre of mass during cornering. The vertical distance between the centre of mass and the roll centre multiplied by the centripetal force determines the **roll couple**. The double wishbone suspension provides great flexibility in the location of the roll centre. By changing the height of the chassis connection points, and hence the inclination of the wishbones, the position of the roll centre can range from below ground to above the centre of mass. The following points are worth noting:

- If the roll centre coincides with the centre of sprung mass there is no roll couple and hence no roll movement when cornering.
- If the roll centre is above the centre of sprung mass, the chassis would roll the 'wrong' way, i.e. the top of the car would lean-in to the inside of corners.

- As the roll centre is lowered the magnitude of roll couple and hence roll movement increases.
- Although it may seem advisable to reduce roll by using a high roll centre, this can introduce two new problems. Firstly, particularly when combined with short swing arm lengths, it produces higher **lateral wheel scrub**. This is the tendency for the wheel contact patch to displace sideways, particularly during bump and rebound. This upsets lateral grip and 'arching' across the two wheels makes it troublesome to set-up ride height in the pits. Secondly, with higher roll centres, the phenomenon of **jacking** increases, whereby the car rises up (rebounds) during hard cornering. We will look at how to quantify jacking later. Also roll is important for tuning the understeer/oversteer balance of the car. For example, if there was no roll, anti-roll bars could not be used as tuning devices.
- A roll centre which actually moves vertically through the ground plane during cornering should be avoided as it will be subject to infinite lateral movement swings to each side of the car as a result of badly conditioned triangles. It is unclear what this would do to the driving experience, but it is certain to unsettle the car.
- Roll centres are often at different heights at each end of the car. A line joining the two roll centres is the **roll axis** – *Figure 3.5*. Some designers believe it is good practice for the roll axis to be inclined so that the roll couple at each end of the car is equalised. This obviously requires the car to be divided lengthwise into two sections and the centre-of-mass height and magnitude evaluated for each end. The motivation is to reduce the need for the chassis to transmit torsional loads, and also as an aid to achieving a good neutral handling balance, however, as we shall see later, there are other ways of achieving this.

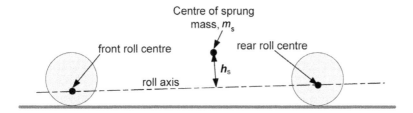

Figure 3.5
The roll axis and roll couple

We will now review the case considered in *Figure 3.3*, i.e. 50 mm bump and 4° roll, and see if converging wishbones perform better than parallel ones. The example uses a static roll centre height of 80 mm and swing arm length of 1.65 m which is about the track width of the car. It can be seen in *Figure 3.6a* that, for a converging wishbone linkage, 4.0° of roll now produces only about 2.0° of adverse camber change which is half the previous value. The fact that the wheel rolls less than the car means that the suspension has produced some **camber recovery** – in this case 50% camber recovery. The small arrow emanating from the centre of the car shows the movement of the roll centre

Figure 3.6a
Converging equal-length wishbones in roll

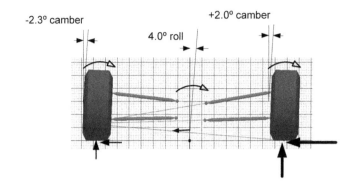

Figure 3.6b
Converging equal-length wishbones in bump

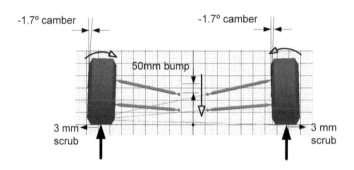

over the 4.0° of roll. This is −8 mm vertical and 142 mm horizontal. The latter figure is slightly high.

Figure 3.6b shows the effect of 50 mm bump. It can be seen that instead of the previous zero camber change there is now −1.7° on both wheels. With 50 mm of rebound this figure is roughly reversed, i.e. +1.7°. In bump/rebound, longer links produce smaller angle changes and hence are desirable. It is also shown that, in this case, there is 3 mm of lateral tyre scrub at 50 mm bump. This is acceptable, but could be reduced if the roll centre was lowered or the swing arm lengthened.

We can conclude that, compared to parallel links, converging links produce better camber control in roll but inferior camber control in bump/rebound. This transition is entirely dependent upon the degree of convergence. In the extreme convergence case, where the top and bottom links meet at a common pivot point on the centre-line of the car, camber change during roll would be zero (100% camber recovery) but camber change during bump/droop would be large. It is effectively the complete opposite of the parallel wishbone case.

3.3.3 Converging unequal-length double wishbones

We will now investigate the effect of adding unequal-length wishbones to the previous converging case. As before, the static roll centre height is 80 mm and swing arm length 1.65 m. The upper wishbone has been shortened and

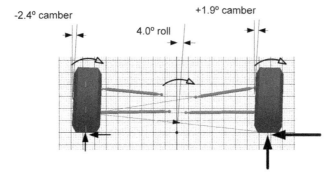

Figure 3.7a
Converging unequal-length wishbones in roll

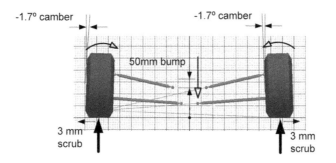

Figure 3.7b
Converging unequal-length wishbones in bump

the lower wishbone lengthened so that the top is about 80% of the bottom. It can be seen from the results presented in *Figures 3.7a* and *b* that this has made virtually no difference to camber change or scrub. (Using unequal-length links can aid camber recovery when the wishbones are short.) Movement of the roll centre is much reduced to only 1 mm vertical and 14 mm horizontal. It can be concluded that the main advantage of unequal-length links is the ability to dial-out roll centre movement.

If this suspension geometry were to be adopted it would probably be associated with a static camber setting of at least −2° to ensure that the heavily loaded outside wheel never approached positive camber.

3.3.4 Summary of suspension front-view requirements

The requirements of a racing suspension can be summarised as follows:

- **Maintain good control of camber during roll** – This means that consideration should be given to camber recovery via converging wishbones and limiting the amount of roll with the use of an ***anti-roll system*** – see *Chapter 4*. Enough static negative camber should be added to the wheels to prevent the heavily loaded outer wheel from going into positive camber at the cornering limit.
- **Provide a stable roll-centre position** – Some wishbone geometries generate large movements of the roll-centre during cornering which can

upset the understeer/oversteer balance of the car. Vertical movement is particularly significant and should be minimised. Reducing the length of the upper wishbone is an effective means of doing this. Avoid a roll-centre that moves through the ground plane.

- **Minimise wheel scrub in bump and rebound** – Lateral displacement of the tyre contact patch should be kept to an acceptable value. This is best limited to movement which can be accommodated by flexing of the tyre wall rather than by dragging the contact patch over the road surface – say a maximum scrub of 5 mm. Scrub is reduced by placing the roll-centre close to ground level.

- **Maintain good control of camber during bump and rebound** – There should not be excessive camber change when either one wheel goes over a bump (single-wheel bump) or both wheels on an axle rise relative to the chassis (two-wheel bump). This requirement is mutually opposed to that of controlling camber during roll. Parallel wishbones are best for controlling camber during bump and converging wishbones best during roll. Maximising the length of wishbones helps however. (The only means of resolving the camber issue in *both* roll and bump is to use an **active suspension** whereby wheel camber is controlled by sensing the camber angle and correcting it with an actuator. Such systems are banned in many forms of racing.)

The effects of modifying suspension geometry on the above requirements are best verified using a suspension design computer package.

3.4 The double-wishbone suspension – side view

The side-view layout of wishbones determines the degree to which various anti-geometries are introduced. By incorporating **anti-dive**, **anti-lift** and **anti-squat** geometries it is possible to prevent some, or all, of the pitching that occurs during hard braking and acceleration. It is important to realise that this does not affect the amount of longitudinal load transfer, but it does change the way that the suspension reacts to longitudinal load transfer. Pitching uses up valuable suspension travel and ground clearance and often adversely affects wheel camber. However most designers advocate restricting the amount of anti-geometry to modest levels – say a limit of 20–30%. The reasons for this are:

- Aggressive anti-dive geometry causes the front suspension to stiffen when encountering a bump under braking and in extremis this can induce tyre **tramp** or bouncing. This is particularly undesirable on road cars as ride comfort is compromised, and on racing cars it can adversely affect grip.

- On the whole, drivers argue that large amounts of anti-geometry make the car feel 'dead' and less responsive.

For the case where relatively highly converging wishbones are used (hence poor camber control in bump), and racing is extensively on smooth circuits, it can be argued that more anti-geometry is desirable. On the other hand, with parallel wishbones, zero may be adopted.

3.4.1 Anti-dive geometry

We saw in *Chapter 1* that braking causes load to be transferred from the rear of the car to the front, normally causing the nose to dive as this additional load compresses the springs. The rear of the car also lifts as the springs at that end are relieved of load. The principle behind anti-geometries is very simple – the suspension link pivots on the chassis are inclined in such a way that the horizontal traction forces counteract the dive forces.

Under maximum braking the *changes* to the front wheel loads, compared to the static case, are the horizontal braking force, F, and the longitudinal weight transfer, ΔW_x. From *section 1.6*:

$$\text{Braking force, } F = W\mu$$

$$\text{Longitudinal weight transfer, } \Delta W_x = \pm \frac{W\mu h_m}{L}$$

Considering *Figure 3.8a*, it can be seen that, if the wishbone pivots on the chassis are parallel to the ground, the wheel is constrained to move perpendicular to the ground. Under these circumstances the horizontal braking force component has no effect on suspension movement and hence there is zero anti-dive.

In *Figure 3.8b* the wishbone pivots have been inclined so that they converge to the rear. Lines are drawn through the upper and lower bearings at the wheel upright parallel to the pivots on the chassis. These two lines intersect at the **side-view instant centre** and the wheel rotates about this point. In effect this has created a **side-view swing axle** of length L_{svsa}. For the normal case of outboard brakes the *change* in forces owing to braking are applied at the tyre contact patch. By taking moments about the instant centre it can be seen that the weight transfer force produces a clockwise moment causing dive. However this is now opposed by an anticlockwise moment from the brake force.

Using *equation [1.10]*

$$\text{Clockwise moment causing dive} = \Delta W_x \times L_{svsa} = \frac{W\mu h_m}{L} \times L_{svsa}$$

$$\text{Anticlockwise moment opposing dive} = W_F\mu \times H_{svsa}$$

Figure 3.8a
Zero anti-dive geometry

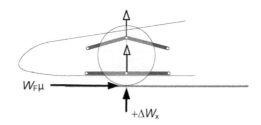

Figure 3.8b
100% anti-dive geometry

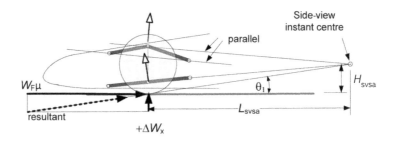

Figure 3.8c
Alternative anti-dive geometry

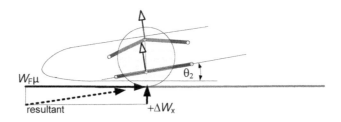

If the two moments are equal and opposite there is no movement of the suspension and this represents 100% anti-dive. (Another view of 100% anti-dive is also shown in *Figure 3.8b* where it can be seen that the resultant force from combining the brake force and the weight transfer force will pass through the side-view instant centre.)

The percentage of anti-dive is reduced as the side-view instant centre height is reduced.

Divide the above moments:

$$\% \text{ anti-dive} = \frac{W_F \mu H_{svsa}}{\dfrac{W \mu h_m L_{svsa}}{L}} \times 100\%$$

$$= \frac{L W_F H_{svsa}}{W h_m L_{svsa}} \times 100\%$$

But % of total braking force at the front, $F_{F\%} = \dfrac{W_F}{W} \times 100\%$

Substituting
$$\% \text{ anti-dive} = \frac{F_{F\%}LH_{svsa}}{h_m L_{svsa}}$$

$$= \frac{F_{F\%}L\tan\theta_1}{h_m} \quad [3.1]$$

where
- $F_{F\%}$ = % of total brake force at front
- L = wheelbase
- H_{svsa} = height of side-view swing axle
- h_m = height of centre of mass
- L_{svsa} = length of side-view swing axle
- θ_1 = slope from contact patch to side-view instant centre.

Figure 3.8c shows an alternative means of obtaining anti-dive. Here the wishbones are parallel but inclined. 100% anti-dive is achieved when the angle of inclination equals the slope of the resultant force. When the angle of inclination is less than this:

$$\text{Slope of resultant} = \tan^{-1}\frac{\frac{W\mu h_m}{L}}{W_F\mu}$$

$$= \tan^{-1}\frac{Wh_m}{LW_F}$$

$$\% \text{ anti-drive} = \frac{\tan\theta_2}{\frac{Wh_m}{LW_F}} \times 100\%$$

$$= \frac{F_{F\%}L\tan\theta_2}{h_m} \quad [3.2]$$

where
- θ_2 = slope between wishbone pivots and ground plane.

3.4.2 Anti-lift geometry

At the rear of the car, longitudinal weight transfer reduces the vertical wheel load causing the springs to relax and the back of the car to lift. *Figure 3.9* shows anti-lift geometry. *Equations [3.1]* and *[3.2]* apply subject to the substitution of the rear brake force percentage, $F_{R\%}$.

3.4.3 Anti-squat geometry

Acceleration causes rearwards longitudinal weight transfer and hence opposite movements to braking. The front of the car will rise and the rear of the car

Figure 3.9
100% anti-lift geometry

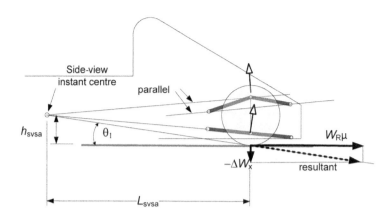

will **squat**. Anti-squat geometry for a rear wheel drive car is similar to brake anti-lift but with one important difference. With an independent suspension (such as double wishbones) the drive torque is resisted by the chassis and the only load path between wheel forces and wishbones is via the wheel hub bearings. Hence the forces must be applied at the wheel centre, as shown in *Figure 3.10*. (This is also the situation with inboard brakes, such as braking on the differential.) Again *equations [3.1]* and *[3.2]* apply, however this time subject to modification of the brake force % term, as all the traction occurs at the rear, and the substitution of θ_3 in *equation [3.1]*.

$$\% \text{ anti-squat} = \frac{L \tan \theta_3}{h_m} \times 100 \qquad [3.3]$$

where θ_3 = slope from wheel centre to side-view instant centre

or, as before:

by alternative method
$$\% \text{ anti-squat} = \frac{L \tan \theta_2}{h_m} \times 100 \qquad [3.4]$$

It is interesting to note that the above formulae are purely geometric and independent of both traction force and coefficient of friction. It follows from the above that, if the instant centre lies at a height between the ground and the wheel centre, there will be positive anti-lift during braking but negative

Figure 3.10
100% anti-squat geometry

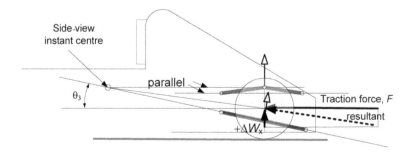

anti-squat during acceleration. Drag racing cars are sometimes designed with over 100% anti-squat (known as **pro-lift**) principally to cause the centre of mass of the car to rise and hence induce more rearward longitudinal load transfer for enhanced rear grip.

It is not possible to remove any lift from the front of a rear wheel drive car in acceleration as there is no traction force available to counter the weight transfer.

EXAMPLE 3.1

Figure 3.11 shows wheel loads under braking obtained from *Example 1.4*. The front wishbones are inclined to form a side-view swing axle with its instant centre having a length of 2100 mm and a height of 105 mm as shown. The rear wishbones are parallel but inclined at 2.5° to the ground.

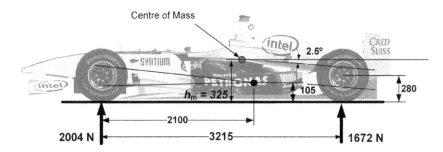

Figure 3.11
Calculating anti-geometries

Determine the percentage of front anti-dive and rear anti-lift under braking, and rear anti-squat under acceleration.

Assume percentage braking at front and rear is proportional to wheel loads.

$$F_{F\%} = \frac{2004}{2004 + 1672} \times 100\% = 54.5\%$$

$$F_{RF\%} = 100 - 54.5 = 45.5\%$$

$$\theta_1 = \tan^{-1}\frac{105}{2100} = 2.9°$$

From *equation [3.1]*   % anti-dive $= \dfrac{F_{F\%} L \tan \theta_1}{h_m}$

$$= \frac{54.5 \times 3215 \times \tan 2.9°}{325} = \mathbf{27.3\%}$$

From *equation [3.2]*   % anti-lift $= \dfrac{F_{R\%} L \tan \theta_2}{h_m}$

$$= \frac{45.5 \times 3215 \times \tan 2.5°}{325} = \mathbf{19.6\%}$$

From *equation [3.4]*   % anti-squat $= \dfrac{L \tan \theta_2}{h_m} \times 100$

$$= \frac{3215 \times \tan 2.5°}{325} \times 100 = 43.2\%$$

3.5 The double-wishbone suspension – top view

The kinematic behaviour of a double-wishbone suspension is not particularly sensitive to plan geometry layout. Consequently it is common to aim the wishbones at strong points on the chassis. Two factors should be considered however:

1. The highest load on the front wishbones often occurs during braking and, for this load case, the force in individual members reduces as the **spread** increases – see *Figure 3.12*. Therefore to remove the need for large wishbone members and to reduce chassis forces, it is advantageous to use as large a spread as possible (but obviously avoiding the problem of a clash between the wishbone and the wheel rim at full steering lock).
2. If the chassis pivots are at some angle to the longitudinal centre-line of the car, as shown in *Figure 3.11*, the main effect is that, when the

Figure 3.12
Wishbones in top view

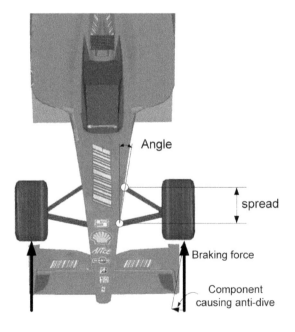

Chapter 3 **Suspension links**

roll centre is above ground level, some anti-dive is introduced. This is because, as shown in *Figure 3.12*, the brake force has a small component perpendicular to the wishbone pivot axis that will oppose longitudinal weight transfer.

3.6 Wishbone stress analysis

Ideally all structural elements of a racing car should be optimised and justified using stress analysis. This section details how design calculations are carried out for wishbone members.

3.6.1 Wishbone loads

Analysis begins by evaluating the wheel loads at the contact patches for various load cases. For the front suspension the critical load cases are likely to be maximum vertical load, braking and cornering. For the rear suspension we can add maximum acceleration. We saw in *Chapter 1* how static wheel loads are changed by lateral and longitudinal load transfer as a racing car accelerates, brakes and corners. It is clear that aerodynamic downforce can increase all of these loads significantly and so must be taken into account. In *Chapter 2* we saw how loads should be subject to a dynamic multiplication factor to account for shock loading. The best way to illustrate the process is to consider an example.

EXAMPLE 3.2

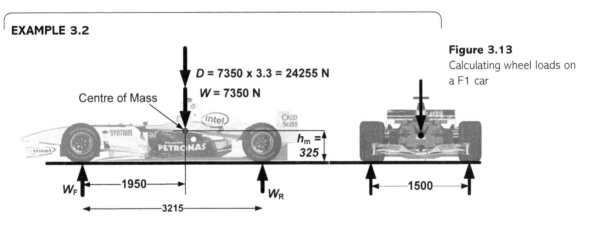

Figure 3.13
Calculating wheel loads on a F1 car

For the peak downforce car used in *Example 1.6* and shown again in *Figure 3.13*, evaluate the design vertical and horizontal loads for the following cases:

(a) front wheel – maximum vertical load,
(b) front wheel – maximum braking,
(c) front wheel – maximum cornering.

Assume a coefficient of friction of 1.2* in (b) and (c).

And by referring to *Example 1.2*:

(d) rear wheel – maximum acceleration at zero downforce.

Assume a coefficient of friction of 1.5* in (d).

* The difference in friction values is due to tyre sensitivity.

(a) Maximum vertical load

The appropriate loads are those at maximum downforce. Applying a dynamic multiplication factor of 3.0 for the portion of vertical load derived from mass and a factor of 1.3 for that derived from aerodynamic downforce gives:

Design vertical load/side $= 0.5 \times [(7350 \times 3) + (7350 \times 3.3 \times 1.3)]$

$= 26\,790$ N

Front wheel design vertical load, $W_{vert} = 26\,790 \times (3215 - 1950)/3215$

$= \mathbf{10\,541\ N}$

(b) Maximum braking

Consider *Example 1.6* which covers the wheel loads under braking at maximum speed:

Front wheel loads (vertical) $= 8135$ N

This must now be increased by the dynamic multiplication factor:

Design load (vertical), W_{vert} $= 8135 \times 1.3$ $= \mathbf{10\,576\ N}$

This is now multiplied by the coefficient of friction, μ, to get the braking force:

Design brake force (longitudinal), $W_{long} = 10\,576 \times 1.2$ $= \mathbf{12\,691\ N}$

(c) Maximum cornering

Effective weight of car, $W = 7350 + (7350 \times 3.3) = 31\,605$ N

From *equation [1.12]*

Maximum cornering force, $F = W \times \mu$ $= 31\,605 \times 1.2$

$= \mathbf{37\,926\ N}$

From *equation [1.13]*

Total lateral weight transfer, $\Delta W_y = \pm \dfrac{Fh_m}{T} = \pm \dfrac{37\,926 \times 325}{1500}$

$= \mathbf{\pm 8217\ N}$

The above figure of 8217 N is the total lateral load transfer for the whole car. Calculating the amount transferred to individual wheels is complex and covered in *Chapter 7*, however for the purpose of calculating suspension forces it is prudent to assume that, say, 62.5% of this figure transfers to the wheel under consideration. That is:

Front outer vertical wheel load
$$= (0.5 \times 31\,605 \times (3215 - 1950)/3215) + (8217 \times 0.625)$$
$$= 11\,354 \text{ N}$$

Apply dynamic multiplication factor, W_{vert} = 11 354 × 1.3 = **14 760 N**

Front outer design cornering force, W_{lat} = 14 760 × μ
= 14 760 × 1.2
= **17 712 N**

(d) Maximum acceleration

It is assumed that the critical case is 'acceleration off-the-line', i.e. negligible downforce and traction-limited acceleration. This is the case considered in *Example 1.3*.

From *Example 1.3*   Rear wheel load = 2628 N

This must now be increased by the dynamic multiplication factor:

Acceleration design load, W_{vert} = 2628 × 1.3 = **3416 N**

This is now multiplied by the coefficient of friction, μ, to get the acceleration force:

Design acceleration force, W_{long} = 3416 × 1.5 = **5124 N**

Note that, as with anti-squat, the acceleration force is applied to the wishbones via the wheel hub bearings and hence occurs at the mid-height of the rear wheel instead of at the tyre contact patch.

Answer – see *Figures 3.14a–d*.

3.6.2 Estimate of wishbone forces from hand calculation and drawing

It is a relatively easy matter to determine the forces in the suspension using simple hand calculations and drawing. It is assumed the individual wishbone members are connected nodally, i.e. the member centroids intersect at the spherical bearing joints in the wheel. This means the forces in the members can be assumed to be either pure tension or compression. Also, at this stage, subtleties in suspension geometry such as caster, scrub radius and camber are ignored as they make little difference to the result.

Figures 3.14a–d
Design wheel loads
– Example 3.2

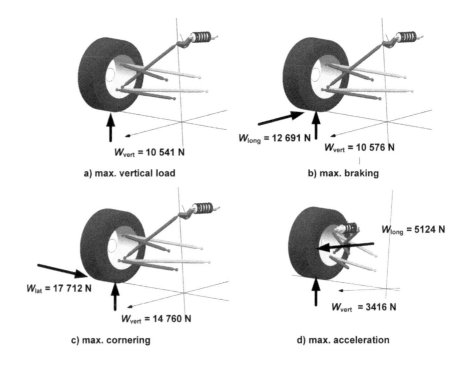

a) max. vertical load — W_{vert} = 10 541 N

b) max. braking — W_{long} = 12 691 N, W_{vert} = 10 576 N

c) max. cornering — W_{lat} = 17 712 N, W_{vert} = 14 760 N

d) max. acceleration — W_{long} = 5124 N, W_{vert} = 3416 N

The suspension links are analysed as a simple pin-jointed space-frame.

Maximum vertical load case

All of the vertical load is resisted by the pushrod and the bottom wishbone – see *Figure 3.15*. Assuming the bottom wishbone is close to horizontal, the vertical component of force in the pushrod must equal the vertical wheel load. If the pushrod is inclined at $\theta°$ to the ground-plane:

$$\text{Vert. wheel load, } W_{vert} = F_{pushrod} \times \sin\theta$$

$$F_{pushrod} = \frac{W_{vert}}{\sin\theta} \text{ N}$$

The horizontal component of force in the pushrod is resisted by the bottom wishbone:

$$\text{Horiz. component in pushrod, } H_{pushrod} = F_{pushrod} \times \cos\theta$$

The forces in the bottom wishbone members can be determined by simply resolving forces at the node, however the easiest method is to draw a **triangle of forces**. A vector is drawn representing the magnitude and direction of the force. Lines are then drawn parallel to the wishbone members to form a triangle and the force magnitudes scaled off as shown in *Figure 3.15*.

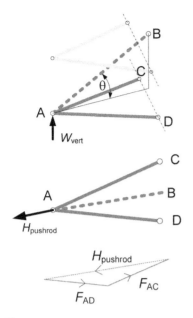
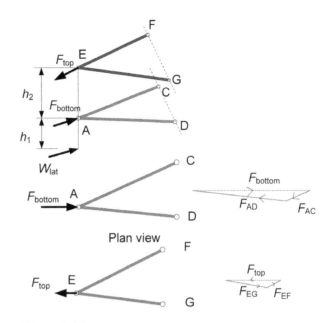

Figure 3.15
Hand analysis of maximum vertical load case

Figure 3.16
Hand analysis of maximum cornering case – horizontal component

Maximum cornering case

The vertical load component is considered as above. For the horizontal component it is necessary to evaluate the equivalent loads on the wishbone nodes. From *Figure 3.16*:

Moments about node A:
$$F_{top} \times h_2 = W_{lat} \times h_1$$

$$F_{top} = \frac{h_1}{h_2} \times W_{lat}$$

If the top wishbone is inclined at $\alpha°$ to the ground-plane, the force in the plane of the wishbone, $F_{wishbone}$, becomes:

$$F_{wishbone} = \frac{F_{top}}{\cos \alpha}$$

Moments about node E:
$$F_{bottom} \times h_2 = W_{lat} \times (h_2 + h_1)$$

$$F_{bottom} = \frac{h_1 + h_2}{h_2} \times W_{lat}$$

A triangle of forces can then be drawn for each wishbone to determine the member forces resulting from the horizontal load. The total load in each

Race car design

Figure 3.17
Hand analysis of maximum braking case – horizontal component

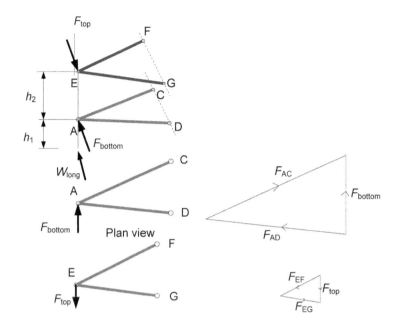

member of the bottom wishbone is obtained by adding the forces from both the vertical and horizontal cases.

Maximum braking case

The forces are applied in the longitudinal direction as shown in *Figure 3.17*.

Maximum acceleration case

As shown in *Figure 3.14d* the acceleration force is applied at the mid-height of the wheel rather than at the contact patch. The longitudinal acceleration force is therefore simply divided between the top and bottom wishbone nodes.

EXAMPLE 3.3

Figure 3.18 shows the dimensions of a wishbone suspension together with maximum braking loads. Determine the forces in the pushrod and wishbone members.

$$F_{pushrod} = \frac{W_{vert}}{\sin \theta} \text{ N}$$

$$= \frac{4500}{\sin 37°} = 7477 \text{ N compression}$$

Horiz. component in pushrod, $H_{pushrod} = F_{pushrod} \times \cos \theta$

$$= 7477 \times \cos 37° = 5971 \text{ N}$$

Chapter 3 **Suspension links**

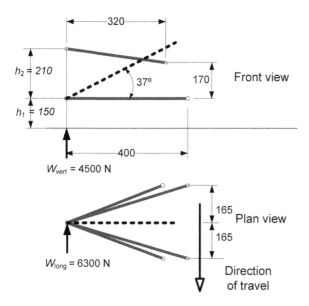

Figure 3.18
Wishbone dimensions and wheel forces for maximum braking

For horizontal load, W_{long}:

$$F_{top} = \frac{h_1}{h_2} \times W_{long}$$

$$= \frac{150}{210} \times 6300 \qquad = 4500 \text{ N}$$

$$F_{bottom} = \frac{h_1 + h_2}{h_2} \times W_{long}$$

$$= \frac{150 + 210}{210} \times 6300 = 10\,800 \text{ N}$$

The force vector diagrams are shown in *Figure 3.19*.

Forces shown positive are tension and those shown negative are compression. It is generally an easy matter to decide if the wishbone member is loaded in tension or compression, however if you are unsure the following procedure can be applied:

1. Add an arrow to the force vector in the same direction as the applied load.
2. Proceed around the force triangle adding arrows so that they form a clockwise or anticlockwise loop.
3. Transfer the arrows to the member drawing adjacent to the node where the load is applied.
4. Add arrows in the opposite direction at the other end of the members.
5. If the arrows on a member 'push' apart the member is in compression.
6. If the arrows 'pull' together the member is in tension.

Race car design

Figure 3.19
Wishbone forces for maximum braking

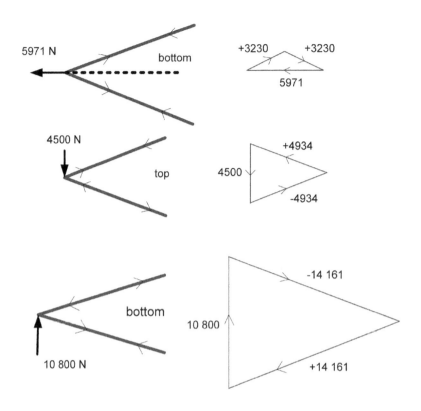

Answer:
Force in pushrod = −7477 N compression
Force in top front = −4934 N compression
Force in top rear = +4934 N tension

Combining the vertical and horizontal load cases for the bottom wishbone:
Force in bottom front = 3230 + 14 161 = +17 381 N tension
Force in bottom rear = 3220 − 14 161 = −10 931 N compression

3.6.3 Computer analysis

Plate 5 shows the output from the finite element package LISA (*ref. 12*) for the same problem as that in *Example 3.3*. Here the wishbones were considered as pin-ended members and a stiff dummy member was introduced to represent the wheel and upright. It can be seen that the results compare closely with those obtained from the hand calculations.

3.6.4 Wishbone sizes

At the highest levels of racing, wishbones are made from carbon composite, otherwise steel is used. Suspension wishbones and pushrods are relatively highly loaded members and consequently it is prudent to use good-quality seamless tubing with a minimum yield stress of 350 N/mm². This is generally in either circular tube form or streamlined elliptical tubing. The design procedures are as described in *section 2.7*. Streamlined compression members are clearly much more likely to buckle about their weak axis and hence the minimum value of second moment of area, I, must be used in the Euler buckling formula – *equation [2.2]*. The properties of some typical wishbone tubes are listed in *Appendix 2*.

EXAMPLE 3.4

For the maximum braking case in *Example 3.3* determine the sizes of suitable streamlined elliptical steel tubing for (a) the front bottom wishbone and (b) the rear bottom wishbone. Assume the steel has a yield stress of 350 N/mm² and an elastic modulus of 200 000 N/mm². Use the sizes given in *Appendix 2*.

Solution
(a) Front bottom wishbone

$$\text{Load in bottom front} = 17\,381 \text{ N tension}$$

From *equation [2.1]*

$$\text{Min. area, } A = \frac{1.5 \times F_t}{\sigma_y} = \frac{1.5 \times 17\,381}{350}$$

$$= 74.5 \text{ mm}^2$$

From *Appendix 2*

Use 28 × 12 × 1.5 elliptical tube   ($A = 87.2 \text{ mm}^2$)

(b) Rear bottom wishbone

$$\text{Load in bottom rear} = -10\,931 \text{ N compression}$$
$$\text{Length of the bottom wishbone member} = \sqrt{(400^2 + 165^2)} = 432.7 \text{ mm}$$

Although the member is welded to the front member at the outer node, it is prudent to assume that the effective length for compressive buckling is the distance between nodes = 432.7 mm as above.

From *equation [2.3]*

$$\text{Euler buckling load, } P_E = \frac{\pi^2 EI}{1.5L^2}$$

$$\therefore \text{Required } I = \frac{P_E \times 1.5L^2}{\pi^2 E}$$

$$= \frac{10\,931 \times 1.5 \times 432.7^2}{\pi^2 \times 200\,000} = 1555.2 \text{ mm}^4$$

From *Appendix 2*
Use 32 × 15.7 × 1.5 elliptical tube ($I = 3163 \text{ mm}^2$)

3.7 Suspension case studies

In the following examples, apart from the Seward F1010, the author has estimated the suspension link geometries from the accompanying photographs for analysis in SusProg. Hence the results are approximate but illustrate the principles of each design philosophy.

3.7.1 F1 car

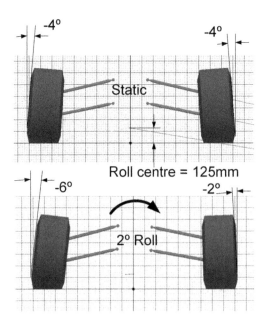

Figure 3.20
F1 car front suspension (reproduced with kind permission from Caterham F1 Team)

Figure 3.20 shows the front suspension of a typical modern F1 car. The suspension geometry is dominated by the aerodynamic requirements for a high nose and the need to maximise airflow underneath it. Earlier cars had either single or double keels extending below the bodywork where the lower wishbones attached, however the current trend is for 'zero-keel' with the wishbones sloping upwards – known as an **anhedral** suspension. This typically results in a relatively high roll-centre with significant lateral scrub, tyre wear and jacking.

The wishbones are virtually parallel and of similar length which means that during roll both the inner and outer wheels suffer adverse camber changes roughly equal to the amount of roll. This is controlled by significant static camber settings of about −4° which will keep the outer wheel in negative camber throughout, however the lightly loaded inner wheel is made less effective with regard to corner grip. In addition, roll is reduced by a very stiff suspension and the use of an anti-roll system. The need for a very stiff suspension also results from aerodynamics. Firstly, the large amounts of downforce would cause ground clearance problems if the suspension was too soft. Secondly, airflow over wings and other surfaces is disrupted if the bodywork moves too much relative to the ground-plane. The roll stiffness of the suspension is roughly equivalent to the roll stiffness of the pneumatic tyres.

It is possible to generate a low roll-centre with an anhedral suspension by inclining the top wishbone more than the bottom. This creates an instant centre on the outside of the wheel at the same side as the suspension. Although this would improve lateral scrub and jacking it means that the tyre would develop more adverse positive camber in roll (the opposite of camber recovery). F1 designers obviously feel that this is not a price worth paying.

Figure 3.21
Dallara F3 car front suspension

3.7.2 Dallara F3 car

Figure 3.21 shows the front suspension of a Dallara F3 car. It can be seen that the wishbones are close to parallel but that the lower wishbones extend under the bodywork to meet at a common point on the centre-line of the car. The fact that the wishbones are of unequal length means that there is a small amount (15%) of camber recovery.

These cars can run with ground clearance as low as 15 mm at the front which means that they also require a very stiff set-up. This, in turn, restricts the amount of roll, ensuring that the outer wheel stays in negative camber despite less negative static camber than the F1 car.

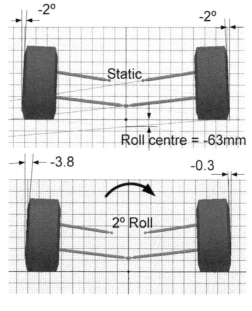

3.7.3 FSAE/Formula Student car

Figure 3.22 shows a typical FSAE/Formula Student car. The tight twisty circuit demands a car with small front and rear tracks. This, combined with regulations that require a relatively wide body, means that the wishbones are comparatively short. A common design response is to adopt highly converging unequal-length wishbones which produce short swing-arm lengths. It can be seen that this results in significant camber recovery with 2° of roll generating only +0.7° of camber change on the loaded wheel. Hence a static camber setting of −1° is sufficient to keep the wheel negative. The arrangement shown has a low roll-centre with good positional stability.

3.7.4 Seward F1010

This is the author's design and contains an extreme wishbone arrangement known as 'Lancaster Links'. *Figure 3.23* shows their form which has ultra-short

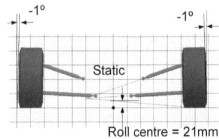

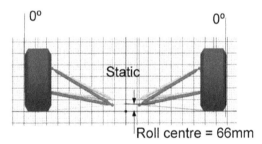

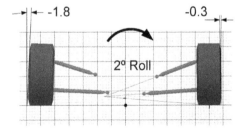

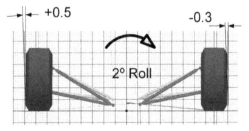

Figure 3.22
FSAE/Formula Student car front suspension (Car from the University of Hertfordshire – 2011)

Figure 3.23
Seward F1010 suspension

swing-arm lengths. Kinematically the wishbones have become a traditional form of suspension known as a **swing axle**. It can be seen that with zero static camber, 2° of roll produces a small amount of negative camber on the outer wheel and positive camber on the inner wheel – both of which are desirable. This represents about 115% camber recovery.

The price that must be paid for this is, unlike with parallel wishbones, considerable camber change in bump and rebound. In fact 25 mm of bump produces over 2° of negative camber. To accommodate this, very robust anti-dive and anti-squat are used. This means that during critical braking and acceleration there is little camber change. Also relatively soft springs mean that, at speed, aerodynamic downforce causes the car to lower by about 12 mm causing a useful –1° of camber throughout. Analysis shows that the much improved control of camber on the lightly loaded inner wheel produces a useful benefit. Despite the fact that there are fewer connections between the suspension members and the chassis, the forces developed at these connections are less than with conventional wishbones, as the force in each of the members is of opposite sign.

> **SUMMARY OF KEY POINTS FROM CHAPTER 3**
> 1. Various forms of the double-wishbone suspension have become almost universally adopted for racing cars.
> 2. For peak cornering grip the top of the wheels should lean towards the centre of rotation of the corner. This implies about −1° camber for the heavily loaded outer wheel and positive camber for the inner wheel.
> 3. Parallel wishbones produce adverse camber changes to all wheels in roll. Converging wishbones and unequal-length wishbones can ameliorate this by producing some camber recovery. Also static camber can be added. This improves the outer wheel but makes the inner wheel worse.
> 4. The requirements of a racing suspension can be summarised as:
>   - Maintain good control of camber during roll*
>   - Provide a stable roll-centre position – particularly vertically
>   - Minimise wheel scrub in bump
>   - Maintain good control of camber during bump (single and two-wheel)*.
> 5. The side-view geometry of wishbones can induce anti-dive, anti-lift and anti-squat effects to reduce pitching of the car during acceleration and braking.
> 6. To evaluate the structural loads in wishbone members it is first necessary to find the design wheel loads for various load cases such as maximum braking and maximum cornering.
> 7. Loads in the wishbone members can be calculated by either simple hand methods and drawing or by using a computer.
> 8. Wishbone member sizes are determined by designing them as either simple struts or ties.

---

* For a passive suspension these two are mutually opposed.

4 Springs, dampers and anti-roll

LEARNING OUTCOMES

At the end of this chapter:
- You will be aware of the types of spring/damper arrangement used on racing cars
- You will be able to specify the length and stiffness of suspension springs
- You will know the basic types of racing damper and how to define the optimum characteristics
- You will be able to design an appropriate anti-roll system

4.1 Introduction

The previous chapter dealt with the suspension geometry that determines the movement path or **kinematics** of the wheels of cars in relation to the chassis. This chapter is concerned with velocity and amplitude of this movement as the suspension reacts to changes in load – i.e. **dynamics**. The primary purpose of the spring/damper system on family cars is to provide ride comfort to the occupants over rough roads. The primary function of spring/dampers on racing cars is to optimise contact between the tyre and the road surface in order to maximise grip. Paul Van Valkenburgh (*ref. 29*) says:

> *'The whole idea is to keep the tyres in the firmest possible contact with the road as long as possible.'*

Figure 4.1 shows the main elements of a typical spring/damper system. The vertical wheel load produces a force in the pushrod which is turned through 90° by the bell-crank and is supported by the spring. The damper prevents undue vibration or oscillation of the spring. The anti-roll system links the two bell-cranks together to stiffen the suspension in roll only. Each of these elements will be considered in more detail to enable the designer to define:

- the maximum bump and rebound wheel movement,
- the spring length and stiffness,
- the damper stroke and settings,

Chapter 4 **Springs, dampers and anti-roll**

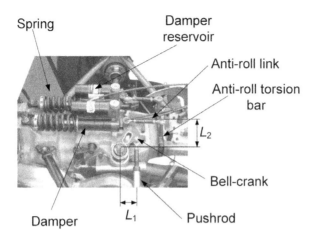

Figure 4.1
Spring/damper components
(Mygale Formula Novis car)

- the bell-crank ratio, L_1/L_2, and hence the **motion ratio** between the **wheel stiffness** and the **spring stiffness**,
- the anti-roll system dimensions.

Figure 4.2 shows three other arrangements for transmitting the load from the wheel to the spring. All of these approaches can be used at the front or rear of the car.

The **external spring/damper** shown in *Figure 4.2a* effectively replaces the pushrod. Although almost universally adopted some years ago, this approach has now been largely replaced by internal arrangements to reduce wind resistance.

The *pullrod* suspension shown in *Figure 4.2b* uses a tension member, in conjunction with the top wishbone to resist vertical wheel loads. As tension members are usually smaller than compression members, this can be a lighter solution with a lower centre of mass.

The *rocker-arm* suspension shown in *Figure 4.2c* replaces the pushrod and the upper wishbone with a double cantilever beam pivoted at the internal node on the chassis. The motion ratio is determined by the length of the two arms. Components in bending tend to be heavier and less stiff than axially loaded members.

Figure 4.2a
External spring/damper
(Formula Jedi)

Figure 4.2b
Pullrod suspension
(Ray Formula Ford)

Figure 4.2c
Rocker-arm suspension
(Van Diemen RF82)

4.2 Springs

In racing the purpose of springs is to enable a degree of independent wheel movement as wheel loads change over irregular road surfaces. The aim is to optimise the grip of the tyre contact patch at each corner of the car. This is part of what can be loosely called **compliance**. Spring movement will also occur as a consequence of changing wheel loads during acceleration, cornering and braking.

The designer needs to define both the **stiffness** and the length of springs. The units of stiffness are N/mm and hence it is a measure of how much the spring moves under a unit load. The following stiffness definitions should be noted:

- ***Spring rate***, K_S, is the stiffness of the actual spring (N/mm). Generally this is a constant and hence most springs have a linear relationship between load and deflection. Linear springs can be used in series with softer **tender springs** to produce a bilinear rate. The tender spring compresses until solid, at which point the combination becomes stiffer. Also **progressive rate springs** exist. These have coils at different spacings. As the spring is compressed a gradually increasing number of coils lock-up, producing increasing stiffness.
- ***Wheel centre rate***, K_W, is the stiffness of the wheel axle in relation to the chassis. It is related to the **spring rate** via the **motion ratio** which is the 'gearing' provided by the mechanism that joins the wheel assembly to the spring connection on the chassis. We will look at this in more detail later.
- **Combined stiffness** or **ride rate**, K_R, is the wheel centre rate, K_W, combined with the **tyre stiffness**, K_T. This represents the effective stiffness of the chassis relative to the road.
- ***Roll rate*** is a measure of the roll couple to cause one degree of roll and is often expressed as the ***roll gradient*** which is the number of degrees of roll per lateral g force. It is dependent upon the ride rate and vehicle track together with any anti-roll system that may be in place.

4.2.1 Hard or soft suspension?

Racing cars are invariably more stiffly sprung than road cars, however this is forced upon designers for the following reasons:

- Racing cars have lower ground clearance and hence less ability to tolerate suspension movement before 'bottoming-out'.
- Aerodynamic downforce causes a racing car to run lower.
- Aerodynamic devices such as wings and floor-pans work best from a relatively stable platform – hence good body control is desirable.
- Higher g forces during acceleration, cornering and braking cause larger load transfers between wheels (however, lower centre of mass on racing cars helps a lot).

Chapter 4 Springs, dampers and anti-roll

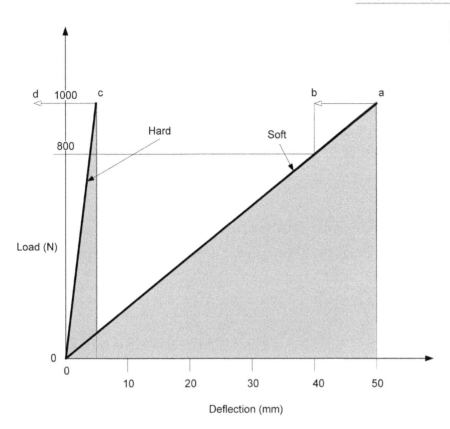

Figure 4.3
Hard versus soft springs

It is undoubtedly easier to put really stiff springs on a racing car which, as well as solving the bottoming-out problem also reduces roll when cornering which, in turn, reduces the issue of adverse wheel camber. However it is true that *to maximise mechanical grip the suspension should be as soft as possible*. This appears to be the opposite of what most manufacturers do when they introduce a sporty version of a road car and hence needs some justification.

Consider *Figure 4.3* which represents the load/deflection graphs for two extreme (but realistic) cases. We will keep the numbers simple and assume that ride rate is the same as spring rate. It is assumed that, when static, the wheel supports a sprung mass of 100 kg which implies a force of about 1000 N in the spring. The 'hard' case assumes a ride rate of 200 N/mm which produces a static deflection of 5 mm. The 'soft' case assumes a ride rate of 20 N/mm which deflects the spring by 50 mm. (The former is similar to a F1 car and the latter to a weekend racer with modest downforce.) Now consider what happens if the wheel passes over a local depression in the road of say 10 mm. The soft case moves from a to b, meaning that the force in the spring drops to 800 N, whereas the hard case moves from c to d implying that the load in the spring is now zero. What we want is for the force in the spring to push the wheel back into contact with the road surface as quickly as possible. Assuming that the unsprung mass of the wheel assembly is 25 kg, the soft spring can provide an average of 900 N to accelerate the wheel downwards.

This equates to an acceleration of $F/m = 900/25 = 36$ m/s^2 = 3.7g. This is in addition to 1g from gravity. Whereas the hard spring runs out of travel after 5 mm and provides no assistance. At best the wheel falls under gravity, at worst it hovers in mid-air until the road surface comes back up to meet it. Another way of looking at the problem is to consider the amount of energy contained in each spring that can perform useful work. This is given by the area under the graphs. The hard spring has $0.5 \times 1000 \times 5/10^3 = 2.5$ Joules whereas the soft spring has ten times as much. The soft suspension is also better over bumps and kerbs as lower upward forces are transmitted to the chassis and are hence less likely to unsettle the chassis or lift other wheels.

4.2.2 Wheel centre rate and natural frequency

As well as determining the amount of suspension movement, the wheel centre rate also influences the natural frequency at which the chassis and wheels vibrate or oscillate. *Figure 4.4* shows a system model for one corner of a car, where:

m_s = corner sprung mass (kg)
m_u = corner unsprung mass (kg)
K_W = wheel centre rate (N/mm)
K_T = tyre stiffness (N/mm).

Figure 4.4
Car corner model

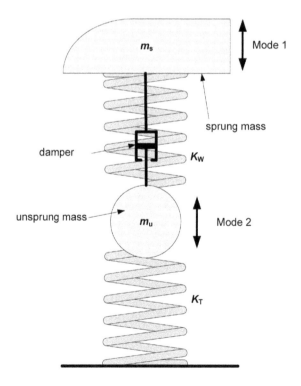

The **sprung mass** consists of the chassis, bodywork, engine, driver etc. The **unsprung mass** is the wheel assembly including tyres, wheels, axles, brakes etc. The mass of components that span between the two, such as suspension links and drive-shafts, is usually included by dividing it between the sprung and unsprung masses.

The system has two significant modes of vibration:

Mode 1 concerns the vibration of the sprung mass of the car relative to the ground. For this case K_W and K_T act as two springs in series, where the **combined stiffness** or **ride rate**, K_R is given by:

$$\frac{1}{K_R} = \frac{1}{K_W} + \frac{1}{K_T} \quad [4.1]$$

Assuming simple harmonic motion, the natural frequency, f_s (Hz), of vibration of the sprung chassis is given by:

$$f_s = \frac{1}{2\pi}\sqrt{\frac{K_R}{m_s}} \quad [4.2]$$

Mode 2 concerns the vibration of the wheel relative to the chassis. In this case K_W and K_T act as two springs in parallel where the combined stiffness is given by ($K_W + K_T$). Because the sprung mass, W_s, is usually big compared to the unsprung mass, W_u, it is usual to assume that the sprung mass is static. The natural frequency of vibration of the unsprung wheel relative to the chassis, f_u (Hz), is given by:

$$f_u = \frac{1}{2\pi}\sqrt{\frac{K_W + K_T}{m_u}} \quad [4.3]$$

Some designers aim for a specific value of the sprung natural frequency, however, particularly when a car has aerodynamic downforce, this is rarely achievable because of ground clearance issues. The frequencies are required for designing the dampers, which is covered later.

EXAMPLE 4.1
Two racing cars each have a corner sprung mass of 110 kg and unsprung mass of 20 kg. The tyre vertical stiffness is 200 N/mm. As described in *section 4.2.1*, one car is 'soft' with a wheel centre rate of 20 N/mm and the other is 'hard' with a wheel centre rate of 200 N/mm.

Determine both the sprung and unsprung natural frequencies for each car.

Soft car

$$\frac{1}{K_R} = \frac{1}{K_W} + \frac{1}{K_T} = \frac{1}{20} + \frac{1}{200} = 0.055$$

Ride rate, K_R = 18.2 N/mm = 18.2×10^3 N/m

$$f_s = \frac{1}{2\pi} \sqrt{\frac{K_R}{m_s}} = \frac{1}{2\pi} \sqrt{\frac{18.2 \times 10^3}{110}} = \textbf{2.05 Hz}$$

$$f_u = \frac{1}{2\pi} \sqrt{\frac{K_W + K_T}{m_u}} = \frac{1}{2\pi} \sqrt{\frac{(20 + 200) \times 10^3}{20}} = \textbf{16.7 Hz}$$

Hard car

$$\frac{1}{K_R} = \frac{1}{K_W} + \frac{1}{K_T} = \frac{1}{200} + \frac{1}{200} = 0.01$$

Ride rate, K_R = 100 N/mm

$$f_s = \frac{1}{2\pi} \sqrt{\frac{K_R}{m_s}} = \frac{1}{2\pi} \sqrt{\frac{100 \times 10^3}{110}} = \textbf{4.8 Hz}$$

$$f_u = \frac{1}{2\pi} \sqrt{\frac{K_W + K_T}{m_u}} = \frac{1}{2\pi} \sqrt{\frac{(200 + 200) \times 10^3}{20}} = \textbf{22.5 Hz}$$

4.2.3 Establishing the required wheel centre rate and roll rate

As already explained, the suspension should be just stiff enough to provide good body control and avoid bottoming out. The designer will generally wish to set the static ride-height at the minimum permitted by the relevant regulations – typically 40 mm. The primary factors that combine to use up this clearance are:

1. Dynamic movement of the chassis owing to road surface roughness, undulations and changes in gradient.
2. Weight transfer from acceleration, braking and cornering. In general, cornering is critical for cars with wide floor-pans and side-skirts whereas braking may be critical for cars with long low noses.
3. Aerodynamic downforce if present.

It can generally be assumed that suspension movement is adequately damped (see later) with the result that, once displaced, the suspension will return to the static position, but will not overshoot significantly or oscillate about the static position. Also the spring/dampers should be fitted with rubber bump stops. These should not be utilised during normal cornering but can be assumed to operate over extreme bumps. The stops effectively make the spring very stiff during the last few millimetres of damper travel.

Three further points should be borne in mind when determining wheel centre rates:

1. If the sprung natural frequencies, f_s, at the front and rear of the car are identical, or very similar, it is likely that the car will suffer from pitching oscillations over some road surfaces. It has been found that if the frequencies differ by at least 10% such oscillations are discouraged.
2. Wheel centre rates at each end of the car influence respective roll rates which determine the amount of lateral load transfer at each end of the car. We have seen from tyre sensitivity that this in turn affects the understeer/oversteer balance of a car. We will see in Chapter 5 how to calculate this effect; however it is often the case that the opposite end of a car to the driven wheels is required to have a higher roll rate for good balance. A lower roll stiffness at the driven end also helps drive traction when accelerating out of corners.
3. Roll rates can be increased with the addition of a suitable anti-roll system without necessarily affecting the vertical wheel centre rate and hence natural frequency.

Knowing the stiffness relationship for springs in series, [4.1]:

$$\frac{1}{K_R} = \frac{1}{K_W} + \frac{1}{K_T}$$

it is an easy matter to find the required wheel centre rate from the ride rate and tyre stiffness:

$$K_W = \frac{K_R K_T}{(K_T - K_R)} \qquad [4.4]$$

The roll rate, K_ϕ, for a car with ride stiffness K_R and track T, is as follows:

$$\text{Roll couple} = C - \text{see Chapter 5, p. 135 for calculation of } C.$$

Resulting lateral load transfer, $F_\phi = C/T$

Vertical displacement of wheel, $\delta_\phi = F_\phi / K_R = C/(T K_R)$

Roll angle, $\theta_\phi = \delta_\phi/(T/2) = 2C/T^2 K_R$ (rad)

$$\text{Roll rate, } K_\phi = C/\theta_\phi = \frac{T^2 K_R}{114.6} \text{ Nm/deg} \qquad [4.5]$$

EXAMPLE 4.2

Estimate the required wheel centre rate for the following two cars, one of which is a relatively low-powered zero downforce car and the other a high downforce car. Also calculate the sprung natural frequencies and roll gradients. Assume both cars have a wide floor-pan and hence there is danger of grounding during cornering roll. For the high downforce car assume the downforce is distributed in the same ratio as the sprung mass. Ignore the effect of any anti-roll system.

	Zero downforce car	High downforce car
Sprung mass, m_s		459 kg
Perp. dist. from m_s centre of mass to roll axis, h_a		220 mm
Ground clearance		40 mm
F:R distribution of sprung mass		40:60
F:R distribution of roll couple		51:49
Front and rear track width, T		1.5 m
Tyre stiffness, K_T		250 N/mm
Downforce	0	6000 N
Lateral g force, G	1.5g	3.0g

Zero downforce car

Allocation of ground clearance budget:

Dynamic movement of the chassis	15 mm
Weight transfer from cornering	25 mm
Total	40 mm

Roll couple, $C = Gm_s h_a = 1.5 \times 9.81 \times 459 \times 0.22 = 1485$ Nm

Front

Roll couple resisted at front, $C_f = 1485 \times 0.51 = 757$ Nm
Resulting weight transfer $= C_f / T = 757/1.5 = 505$ N
Required front ride rate, $K_R = 505/25 = $ **20.2 N/mm**
(where 25 = cornering ground clearance budget)

Required front wheel centre rate:

$$K_W = \frac{K_R K_T}{(K_T - K_R)} = \frac{20.2 \times 250}{(250 - 20.2)}$$

$$= \mathbf{22.0 \text{ N/mm}}$$

Front sprung mass/wheel $= 459 \times 0.4/2 = 91.8$ kg

Front sprung natural frequency:

$$f_s = \frac{1}{2\pi} \sqrt{\frac{K_R}{m_s}} = \frac{1}{2\pi} \sqrt{\frac{20.2 \times 10^3}{91.8}} = \mathbf{2.36 \text{ Hz}}$$

Rear

Roll couple resisted at rear $= 1485 \times 0.49 = 728$ Nm
Resulting weight transfer $= 728/1.5 = 485$ N
Required rear ride rate $= 485/25 = $ **19.4 N/mm**

Required rear wheel centre rate:

$$K_W = \frac{K_R K_T}{(K_T - K_R)} = \frac{19.4 \times 250}{(250 - 19.4)}$$

$$= 21.0 \text{ N/mm}$$

Rear sprung mass/wheel $\quad = 459 \times 0.6/2 \quad = 137.7$ kg

Rear sprung natural frequency:

$$f_s = \frac{1}{2\pi}\sqrt{\frac{K_R}{m_s}} = \frac{1}{2\pi}\sqrt{\frac{19.4 \times 10^3}{137.7}} = \mathbf{1.89 \text{ Hz}}$$

Body roll $\quad = \tan^{-1}(25/750) \quad = 1.9°$
Roll gradient $\quad = 1.9/1.5 \quad = \mathbf{1.27 \text{ deg/g}}$

High downforce car

Allocation of ground clearance budget:

Dynamic movement of the chassis	10 mm
Weight transfer from cornering + downforce	30 mm
Total	40 mm

Roll couple, C $\quad = 3 \times 9.81 \times 459 \times 0.22 = 2970$ Nm

Front

Roll couple resisted at front $\quad = 2970 \times 0.51 \quad = 1515$ Nm

Resulting weight transfer $\quad = 1515/1.5 \quad = 1010$ N
Increased load from downforce $= 6000 \times 0.4/2 \quad = 1200$ N
Total increase $\quad = 2210$ N

Required front ride rate, K_R $\quad = 2210/30 \quad = \mathbf{73.7 \text{ N/mm}}$

Required front wheel centre rate:

$$K_W = \frac{K_R K_T}{(K_T - K_R)} = \frac{73.7 \times 250}{(250 - 73.7)}$$

$$= \mathbf{104.5 \text{ N/mm}}$$

Front sprung mass/wheel $\quad = 459 \times 0.4/2 \quad = 91.8$ kg

Front sprung natural frequency:

$$f_s = \frac{1}{2\pi}\sqrt{\frac{K_R}{m_s}} = \frac{1}{2\pi}\sqrt{\frac{73.7 \times 10^3}{91.8}} = \mathbf{4.51 \text{ Hz}}$$

Rear

Roll couple resisted at rear	= 2970 × 0.49	= 1455 Nm
Resulting weight transfer	= 1455/1.5	= 970 N
Increased load from downforce	= 6000 × 0.6/2	= 1800 N
Total		= 2770 N
Required rear ride rate	= 2770/30	= **92.3 N/mm**

Required rear wheel centre rate:

$$K_W = \frac{K_R K_T}{(K_T - K_R)} = \frac{92.3 \times 250}{(250 - 92.3)}$$

$$= \textbf{146.3 N/mm}$$

Rear sprung mass/wheel = 459 × 0.6/2 = 137.7 kg
Rear sprung natural frequency:

$$f_s = \frac{1}{2\pi}\sqrt{\frac{K_R}{m_s}} = \frac{1}{2\pi}\sqrt{\frac{92.3 \times 10^3}{137.7}} = \textbf{4.12 Hz}$$

Vertical displacement from roll	= 970/92.3	= 10.5 mm
Body roll	= tan^{-1} (10.5/750)	= 0.802°
Roll gradient	= 0.802/3.0	= **0.27 deg/g**

Check using roll rate *equation [4.5]*:

Roll rate	= $\dfrac{T^2 K_R}{114.6}$	= (1.5² × 92.3 × 10³)/114.6
		= 1812 Nm/deg
From above Roll rate		= 1455/0.802
		= 1814 Nm/deg ✓

4.2.4 Specifying springs

Spring rate

Having established an initial estimate of the required wheel centre rate it is now necessary to calculate the resulting spring rate. Geometrically, the simplest form of suspension is the rocker-arm (*Figure 4.2c*) and this will be used to understand the principles. *Figure 4.5* shows the rocker-arm in diagrammatic form.

It can be seen that in this case one arm of the rocker is of length L and the other of length $R_m L$, where R_m is the **motion ratio**. It can be seen from simple

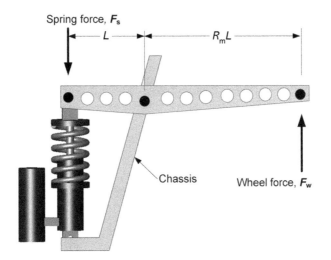

Figure 4.5
The rocker-arm suspension

proportions that if the spring moved a distance of d, the wheel would move a distance of $R_m d$. The motion ratio is taken to be greater than one if the wheel moves more than the spring. (An alternative term is the **installation ratio** which is generally taken as the inverse of the motion ratio.)

Taking moments about the central pivot point:

$$F_S \times L = F_W \times R_m L$$
$$F_S = R_m \times F_W$$

and $\quad F_W = F_S/R_m$

Stiffness or rate is defined as force per unit displacement:

$$\text{Wheel centre rate, } K_W = F_W/R_m d$$
$$\text{Spring rate, } K_S = R_m F_W/d$$

Dividing K_S by K_W gives the **ratio of stiffnesses** $= R_m^2$

i.e $\quad\quad\quad\quad\quad\quad\quad K_S = R_m^2 K_W \quad\quad\quad$ [4.6]

The above is true for all forms of double-wishbone suspension, however determining the motion ratio for pushrod and pullrod suspensions is more difficult and usually requires a careful scale drawing, a physical model or suspension analysis software such as SusProg. In the case of the scale drawing, the procedure is to move the wheel in say 5 mm increments and plot the resulting change in length of the spring.

The benefit of a pushrod or pullrod arrangement is that the spring/damper can be enclosed within the bodywork for aerodynamic reasons. A bell-crank is used to rotate the force from the wheel through a convenient angle for packaging of the spring/damper. The following points should be noted – see *Figure 4.6*.

Figure 4.6
Pushrod – bell-crank arrangement

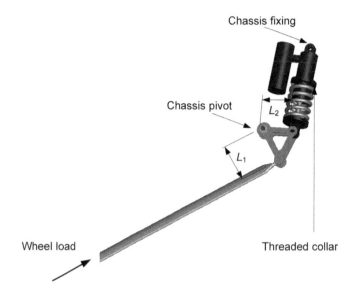

1. The pushrod, bell-crank and spring/damper should lie on the same plane to avoid out-of-plane bending. Normally this is the case under static loading but it could be argued that the components should move onto the same plane at full bump when the pushrod load is at a maximum.
2. Varying the bell-crank ratio, L_1/L_2, is a convenient means of adjusting the motion ratio, R_m.
3. The bell-crank must be of adequate size to produce sufficient spring displacement.
4. Normally the angle between the faces of the bell-crank and the pushrod and the spring/damper are set at right-angles. This produces close to a linear relationship between wheel movement and spring compression – i.e. close to a constant value of the motion ratio, R_m. Sometimes a non-linear relationship is desired. The example shown in *Figure 4.1* is such a case, where, as the bell-crank rotates, L_1 reduces and L_2 increases thus further reducing the motion ratio, R_m, and producing an increasingly stiff suspension with bump. This is known as a **rising rate suspension**.

Spring length

We saw in the previous section that the maximum amount of bump displacement at the wheel is generally about the same as the ground clearance. In addition to bump we need to include the distance that the spring initially compresses when going from zero to the static sprung corner weight. This is given by:

$$\text{Initial compression} = \frac{\text{sprung corner weight}}{\text{wheel centre rate}}$$

then Total wheel movement = bump + initial compression

$$\text{Total spring movement} = \frac{\text{total wheel movement}}{\text{motion ratio, } R_m}$$

The required overall uncompressed spring length needs to be about twice the spring movement to prevent the spring ever becoming solid.

EXAMPLE 4.3

Determine the spring rate and spring length for the front suspension of the zero downforce car in *Example 4.2* assuming a motion ratio, R_m, of 1.3.

Extracting values from *Example 4.2* where appropriate:

Wheel centre rate, K_W = 22.0 N/mm

Spring rate, K_S = $R_m^2 K_W$ = $1.3^2 \times 22.0$

= **37.2 N/mm**

Maximum bump = 40 mm

$$\text{Initial compression} = \frac{\text{sprung corner weight}}{\text{wheel centre rate}} = \frac{91.8 \times 9.81}{22.0}$$

= 41 mm

Total wheel movement = 40 + 41 = 81 mm

$$\text{Total spring movement} = \frac{\text{wheel movement}}{R_m} = \frac{81}{1.3}$$

= 62.3 mm

Minimum spring length = 62.3 × 2 = **125 mm**

Spring pre-load

With the normal spring/damper arrangement, as shown in *Figure 4.6*, the spring is seated on a threaded collar which can be screwed up and down the body of the damper. The default position is to loosely hand-tighten the collar until the spring is retained with the damper fully extended. With the car supported on its wheels the load in the spring will be the static sprung corner weight, and the damper rod will retract into the damper body by an amount equal to the compression of the spring.

If the threaded collar is backed-off, the load in the spring remains the same, but the damper rod will retract further into the cylinder and the ground clearance of the car will be reduced. This is a viable means of adjusting ride height, however it uses up valuable bump damper travel. It may be better to adjust ride height by changing the length of the pushrod/pullrod.

If, on the other hand, the threaded collar is tightened, the load in the spring still remains the same, but the damper rod will protrude further from the cylinder and the ground clearance of the car will be increased. This reduces available rebound travel of the damper but this may not be a disadvantage. It is probable (and desirable) that during maximum roll of the car there will always be a minimum positive load on the inner wheel contact patch. In other words, the inner wheel will never lift-off, and the load in the spring will never reduce to zero. The rebound travel at very low spring loads is therefore never utilised. The threaded collar can thus be tightened by a distance equal to the minimum spring load divided by the spring stiffness. This makes more effective use of the damper stroke by permitting more bump travel.

It is important that full suspension movement during corner roll is not impeded by a damper reaching its travel limit in either direction, as this produces a significant change in stiffness which is likely to cause a sudden upset to the balance of the car.

4.3 Dampers

Dampers are often referred to as **shock absorbers** or simply **shocks**. The principal purpose of dampers is to prevent dynamic oscillations of both the sprung and unsprung masses. Although small damper adjustments may play a role in the fine-tuning of the transient behaviour of a racing car, it should not be used as a tool for steady-state tuning of the understeer/oversteer balance. This is because the effect of damping changes with time and would hence produce different behaviour over short and long corners.

We have already seen that the sprung and unsprung masses have different natural frequencies and therefore different damping requirements. In addition the sprung mass may oscillate vertically or in pitch or roll. Ideally all these situations require individual damping systems, however most cars have to make do with a single damper at each corner. Some compromise is therefore necessary.

If a mass, which is supported by a spring, is displaced from its static equilibrium position and released it will oscillate. In theory, without a damper, the oscillations continue indefinitely at constant amplitude and frequency. In practice, however, natural damping from internal friction and external air resistance causes oscillations to decay over a finite period as shown in *Figure 4.7*.

A damper, connected in parallel with the spring, is used to modify this behaviour. **Viscous dampers** operate by forcing a fluid through an orifice which produces a force that restrains the spring motion. This force is propor-

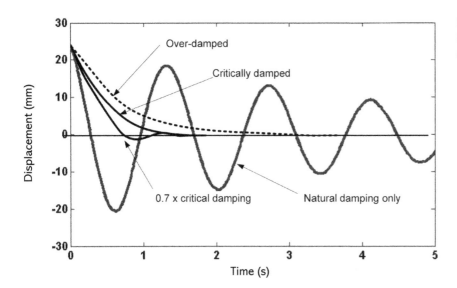

Figure 4.7
Damped oscillation of a spring/mass system

tional to the velocity of motion. The **damping coefficient**, **C**, is a measure of the damping force at a velocity of 1 m/s, hence the units are N/m/s or Ns/m. **Critical damping** is defined as that which causes the mass to return to its steady-state position without any over-shoot – see *Figure 4.7*. The value of the damping coefficient for critical damping of the sprung mass:

$$C_{crit} = 4\pi m_S f_S \quad [4.7]$$

where
m_S = sprung mass (kg)
f_S = natural frequency (Hz).

The above assumes that the damper acts directly between the sprung and unsprung masses. If, as is normally the case, the damper acts between the spring mounts, the motion ratio, R_m, must be taken into account:

$$C_{crit} = 4\pi m_S f_S R_m^2 \quad [4.8]$$

If the damping coefficient is greater than critical the system is said to be **over-damped** and will take a longer time to return to the steady-state position. If the damping coefficient is less than critical the system is **under-damped** and some oscillation about the steady-state position will take place. The ratio of the damping coefficient adopted, **C**, to the critical damping coefficient, **C_{crit}**, is termed the **damping ratio**, ζ (zeta):

$$\zeta = \frac{C}{C_{crit}} \quad [4.9]$$

For ordinary cars a damping ratio of about 0.25 gives the best compromise between comfort (ride) and performance (handling). For racing cars, where comfort is not important, an average damping ratio of about 0.65–0.7 has been found to work well. As can be seen from *Figure 4.7*, this produces only a small amount of overshoot and returns to the neutral position more quickly than critical damping.

EXAMPLE 4.4

(a) A car in *Example 4.1* has a corner sprung mass, m_S, of 110 kg and a sprung natural frequency of 2.05 Hz. Assuming a motion ratio, R_m, of 1.3 determine the value of damping coefficient, C, to produce a damping ratio, ζ, of 0.7.

(b) The same car has an unsprung mass of 20 kg and a natural frequency of 16.7 Hz. If the damping coefficient, C, from part (a) is adopted, what is the unsprung damping ratio?

(a)
$$\zeta = \frac{C}{C_{crit}} = 0.7$$

Damping coefficient, $C = 0.7\, C_{crit} = 0.7 \times 4\pi m_S f_S R_m^2$

$$= 0.7 \times 4\pi \times 110 \times 2.05 \times 1.3^2 = \mathbf{3352\ N/m/s}$$

(b)
$$C_{crit} = 4\pi m_U f_U R_m^2$$

$$= 4\pi \times 20 \times 16.7 \times 1.3^2 = 7093\ \text{N/m/s}$$

Unsprung damping ratio, $\zeta = \dfrac{C}{C_{crit}} = \dfrac{3352}{7093} = \mathbf{0.47}$

Damper characteristics are often illustrated on a force against velocity graph, known as a damping curve, where the damping coefficient is the slope of the line. The value from *Example 4.4* is shown in *Figure 4.9a* (on page 108). It is found, however, that the slopes of the damping curve require further modification for optimum performance. To gain some insight into these modifications it is necessary to consider rough relative velocities of the wheel during bump and rebound. *Figure 4.8* shows a wheel on a car travelling at 28 m/s (100 km/h) which encounters a 10 mm high obstruction such as a low kerb. It can be seen that over a distance of 75 mm the wheel rises by 10 mm:

Upward bump velocity

$$= 28 \times \frac{10}{75} = \mathbf{3.7\ m/s}$$

Now consider the other end of the obstruction. The downward acceleration of the wheel depends upon the force in the spring plus gravity. Assuming an average corner load in the spring of 1000 N and an unsprung wheel mass of 20 kg:

Chapter 4 Springs, dampers and anti-roll

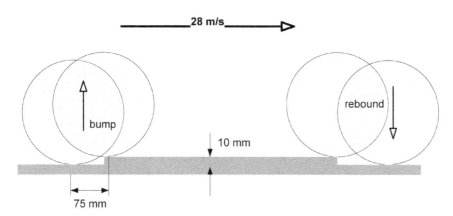

Figure 4.8
Relative wheel velocities during bump and rebound

$$\text{Downward acceleration from spring} = \frac{\text{force}}{\text{mass}} = \frac{1000}{20} = 50 \text{ m/s}^2$$

$$\text{Total downward acceleration, } a = 50 + 9.81 = 60 \text{ m/s}^2$$

From Newton   Velocity = $2 \times a \times$ distance   = $2 \times 60 \times 0.01$

Downward rebound velocity = 1.2 m/s

It can be seen from the above that the bump velocity is more than double that during rebound. It could therefore be thought that the greater bump velocity requires more damping, however the opposite is the case. High velocities produce high damping forces and hence high accelerations, which, when acting upwards, unsettle the sprung chassis and cause large weight transfers between the wheels. It is also suggested that the primary role of bump damping is to control the vibration of the unsprung wheel over bumps, whereas the primary role of rebound damping is to control the motion of the sprung chassis during corner roll. We can therefore conclude that if bump damping is too low **wheel hop** will result and if it is too high it will unsettle the chassis. If rebound damping is too low the chassis will wallow in corners and if it is too high there is a risk of **jacking down** as high rebound damping stops the wheel from returning before a further bump disturbance pushes it back up. A common approach is therefore, as before, to calculate a damping coefficient of $0.7C_{crit}$ based on the sprung mass but to reduce the bump damping coefficient to 2/3 of this value and to increase rebound damping by a factor of 3/2. The modified damping curve is shown in *Figure 4.9b*. The average damping over a full cycle is thus roughly $0.7C_{crit}$.

A final modification is to impose a slow region and a fast region onto the damping curve. This is again an attempt to reduce the impact of large damping forces acting on the sprung mass. It is common practice to halve the value of the damping coefficient for all velocities above a transition point – *Figure 4.9c*. The velocity at which the transition between slow and fast occurs is often as low as 0.05 m/s. The velocity is given by the slope of the curve in *Figure*

Race car design

Figure 4.9a
Initial damping curve

Figure 4.9b
Modified damping curve

Figure 4.9c
Final damping curve

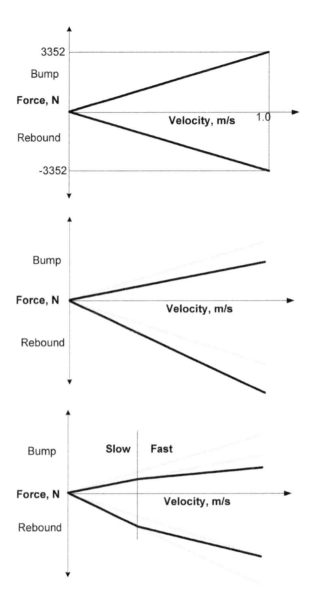

4.7. For simple harmonic motion the maximum velocity occurs as the mass crosses the neutral point and is given by $v_{max} = 2\pi f A$, where A is the amplitude (or size of bump). For a 'soft' car with $f = 2$ Hz a velocity of 0.05 m/s implies a transition amplitude of only about 4 mm. It can be seen from *Figure 4.7* that 0.7 critical damping slows the velocity down by a factor of about two which gives an amplitude of 8 mm. For all bumps greater than this the system would be more under-damped. Also as the frequency rises for stiffly sprung high-downforce cars the transition amplitude reduces.

The values of damping coefficient suggested here are sufficient for the initial specification of dampers but final settings should be based on careful track testing with a knowledgeable driver as described in *Chapter 11*.

> **EXAMPLE 4.5**
>
> Continue *Example 4.4* to establish the final values of the damping coefficients.
>
> Initial damping coefficient, $C = 0.7 C_{crit}$ = 3352 N/m/s
>
> *Low frequency*
>
> Bump (compression) slope = 2/3 × 3352 = 2235 N/m/s
>
> Rebound (tension) slope = 3/2 × 3352 = 5028 N/m/s
>
> *High frequency*
>
> Bump (compression) slope = 1/2 × 2235 = 1118 N/m/s
>
> Rebound (tension) slope = 1/2 × 5028 = 2514 N/m/s
>
> The above damping coefficients give the slopes of the lines in *Figure 4.9c*.

4.3.1 Damper selection

The vast majority of dampers currently used are known as **coilovers**, where the spring is coiled concentrically over the damper as shown in *Figures 4.1* and *4.2*. Two types of coilover damper are available:

1. **Twin-tube dampers**

   Twin-tube dampers consist of a central tube which contains a piston and an outer tube that acts as a reservoir for displaced damper oil. To prevent aeration of the oil during intensive use these dampers may keep the oil under pressure, usually with nitrogen gas. They are generally manufactured from steel tube but aluminium alloy is sometimes available to save weight.

   Twin-tube dampers are relatively cheap and are usually fitted as standard to road cars. Aftermarket versions are available with adjustable valves so that damping coefficients can be varied. Consistent damping requires a reasonable volume of oil to be moved through the damper orifice and the twin-tube design means that the piston diameter is relatively small. Hence they are not ideal for short-stroke or compact applications. Also dampers dissipate energy by heating the oil, and the twin-tube design makes oil cooling less effective. As the oil temperature rises, viscosity, and hence effective damping, reduce.

2. **Mono-tube dampers**

   The simplest form of mono-tube damper consists of a single tube and locates the oil/gas reservoir in series with the piston. For a given stroke length this makes the mono-tube longer than the equivalent twin-tube.

To alleviate this issue twin-tube dampers are available with separate reservoirs, either rigidly mounted piggyback style on the side of the tube, as shown in *Figure 4.5*, or remotely via a hose. They are generally manufactured from aluminium alloy.

Mono-tube dampers are considerably more expensive than twin-tubes. They are available with fixed valving – usually to the designer's specification, or with one-, two- or four-way adjustment. Four-way adjustment provides individual control over the four different damping coefficient slopes shown in *Figure 4.9c*. Mono-tubes, with separate reservoirs, work better for compact applications and have better fluid cooling for more consistent performance throughout a race.

Damper characteristics should be verified by physical testing in a dedicated test rig. The traditional means of doing this is to connect the damper to a rotating flywheel via a crank. The crank is set to move the damper through a specific amplitude, A, and the frequency of rotation, f, is set. For the resulting sinusoidal motion the maximum velocity is

$$v_{max} = 2\pi f A$$

where v_{max} occurs at zero displacement. The test is repeated for a range of frequencies and *Figure 4.10* shows typical results from three tests. The main points of interest are the force readings at zero displacement – indicated as 1, 2 and 3 on *Figure 4.10*. These can be plotted against the corresponding velocities calculated from the above formula to give points on a damping curve such as *Figure 4.9c*. The tests are further repeated with adjustable dampers at different settings. Modern test rigs can provide direct plots of the damper curves.

Figure 4.10
Damper force/displacement curve from traditional test rig

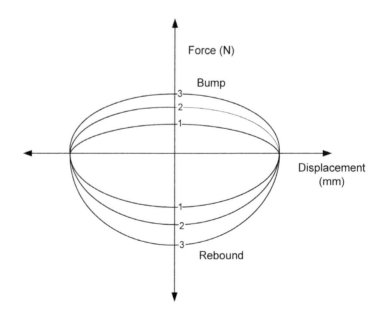

4.4 Anti-roll systems

Anti-roll systems have two main functions:

1. To reduce cornering roll. Roll in itself is not too detrimental on cars which have a low centre of mass; however, with most suspension geometries it produces adverse positive camber change to the wheels which reduces grip. Reducing roll does not reduce total lateral load transfer. Softly sprung cars may require anti-roll systems at both front and rear. Stiff high-downforce cars are unlikely to require additional roll stiffening from an anti-roll system.
2. To act as an adjustable tuning aid by increasing the roll stiffness at one end of the car relative to the other. As previously explained this exploits the phenomenon of tyre sensitivity to affect the handling balance and hence reduce excessive under- or over-steer. Rear wheel drive cars generally require anti-roll systems at the front and vice versa.

There is a clear disadvantage in providing too much roll stiffness through the anti-roll system. With over-stiff systems the suspension ceases to be truly 'independent'. The wheels on one axle are effectively joined together and act like a rigid axle in single-wheel bump. A car is more likely to lift a wheel in hard cornering. Although the stiffness of the suspension is not affected in two-wheel bump and rebound, under single wheel movements the suspension becomes stiffer and hence less effective. For this reason it is recommended that an anti-roll system does not contribute more than 50% of the roll stiffness at either end of the car. Needless to say, the free movement of the suspension throughout its full bump and rebound range should not be impeded by the anti-roll system.

4.4.1 Anti-roll systems

On road cars anti-roll systems usually take the form of a substantial U-shaped torsion bar which is pivoted on the underside of the car and spans between the front or rear wheel uprights. This is usually referred to as an anti-roll bar or **arb**. On racing cars, with inboard suspension components, a smaller and lighter system is possible by using links that join the arms of the torsion bar to the bell-cranks. Examples are shown in *Plate 4* and *Figure 4.1*. The links are connected to a point on the bell-crank that moves approximately parallel to the longitudinal axis of the car. Moving the link connection up or down the arm changes the stiffness of the anti-roll bar and is thus a powerful tuning tool.

Figure 4.11a shows a U-bar anti-roll system in the static, unloaded, position. The bar is free to rotate in the lower pivots which are connected to the chassis. The horizontal links are connected to the bell-cranks.

Figure 4.11b shows the effect of two-wheel rebound (or bump). The bar

Figure 4.11
U-bar anti-roll systems

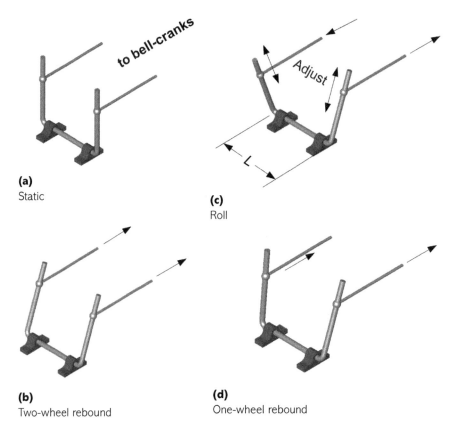

(a) Static

(c) Roll

(b) Two-wheel rebound

(d) One-wheel rebound

simply pivots forwards (or backwards) and is unstressed. The bar thus has no effect on suspension stiffness.

In roll, the motion of the bell-cranks forces the arms of the U-bar in opposite directions – *Figure 4.11c* – which clearly puts the arms in bending and the base of the bar in torsion. The system is a torsion spring whose stiffness must be added to that of the suspension springs. In the example shown, the bell-crank links can be moved up-and-down the U-bar arms in order to vary the torque and hence the stiffness of the system. This is a powerful tuning mechanism to achieve balanced handling of the car.

Under one-wheel rebound the U-bar will stiffen the dropping wheel and transmit some of the downward motion to the other wheel. The two U-bar arms will move in the same direction but by differing amounts depending upon the relative stiffness of the bar – *Figure 4.11d*. The arb has thus made individual wheel movements less independent.

Figure 4.12 shows an alternative form of anti-roll system known as the T-bar. The principle of operation is similar to the U-bar. Under roll, twisting deformation

Figure 4.12
T-bar anti-roll system

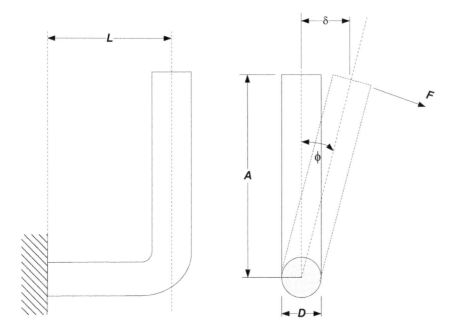

Figure 4.13
U-bar anti-roll stiffness calculations

takes place in the vertical torsion bar as well as bending in the arms. The usual method of providing adjustment is to change the stiffness of the arms either by changing the number of 'blades', like a leaf spring, or by rotating the arms so that the effective second moment of area (and hence bending stiffness) change.

4.4.2 Anti-roll system calculations

It is important to be able to determine the dimensions of an arb to provide the required increase in roll stiffness. Consider the roll case shown in *Figure 4.11c*. Because the arms move in opposite directions it can be seen that, from symmetry, there will be no rotation at the mid-point of the torsion bar. This point can therefore be considered to be fixed as shown in *Figure 4.13* and we consider **half** of the torsion bar to be of length L. The objective is to find the stiffness (N/mm) at the top of the arm of length A.

$$\text{Stiffness} = \frac{F}{\delta} = \frac{F}{A\phi} \quad \text{where } \phi = \text{rotation in radians}$$

Also the torsional rotation of a **solid** circular shaft is given by:

$$\phi = \frac{32LFA}{\pi GD^4}$$

where G = **shear modulus** or **modulus of rigidity**

= 79 300 N/mm² for typical spring steels

Substituting:

$$\text{Stiffness}, K_{bar} = \frac{\pi G D^4}{32 L A^2} \quad [4.10]$$

In addition to the above the arms will bend, although this generally only makes a small difference to the stiffness. The deflection of a cantilever with a point load at the end is given by:

$$\delta = \frac{F A^3}{3 E I} \quad [4.11]$$

where E = **modulus of elasticity**

= 207 000 N/mm² for typical spring steels

I = **second moment of area**

$$= \frac{\pi D^4}{64} \text{ for a solid circular bar}$$

$$\text{Stiffness}, K_{arm} = \frac{F}{\delta} = \frac{3 E I}{A^3} = \frac{3 E \pi D^4}{64 A^3} \quad [4.12]$$

The torsion bar and arm elements are in series where the combined stiffness, K_{arb}, is given by:

$$\frac{1}{K_{arb}} = \frac{1}{K_{bar}} + \frac{1}{K_{arm}} \quad [4.13]$$

Points to note
1. The anti-roll system must be of adequate size to permit the full bump and rebound movement of the suspension.
2. If a hollow tube is used in place of a solid bar, the value of D^4 in the above equations is replaced by $(D^4 - d^4)$ where D is the external diameter and d is the internal diameter.
3. In order to evaluate the effect of the arb on wheel stiffness, K_{arb} must be multiplied by the appropriate motion-ratio squared (R_m^2). This may have a different value from the spring motion ratio depending upon the radius of the connection point on the bell-crank. In this case the motion ratio is defined as the vertical wheel movement divided by the movement of the arb at the point where the link connects to the arm.
4. The addition of an arb to a car will mean that the wheel centre rate (and hence the ride rate) will be higher in roll than it will be in bump and rebound.
5. The arb will provide additional adverse stiffening to single wheel movements as shown in *Figure 4.11d*, however, the effect is less than the

increase in roll stiffness. Firstly, because the two ends of the bar move in the same direction which reduces the arb stiffness to 50% of the value in roll. Secondly, because the arb acts in series with the spring on the opposite side of the car.

6. Anti-roll bars must be manufactured from good-quality spring steel which is likely to have a yield stress of around 1500 N/mm^2. This will be a high-carbon steel which is likely to have very poor weld qualities and also require heat-treatment before and after bending.

7. Care must be taken not to overstress the arb during large roll movements. The critical stress is likely to be the shear stress in the torsion bar which, given spring steel with the above yield stress, should never exceed 0.6 × 1500 = 900 N/mm^2. The maximum shear stress, τ, in a solid circular bar subject to torsion moment, T, is given by:

$$\tau = \frac{16T}{\pi D^3}$$

In our case the value of torsion moment, T, is given by the following – see *Figure 4.12*:

$$T = K_{arb} \times \delta \times A$$

Hence $$\tau = \frac{16 K_{arb} \delta A}{\pi D^3} \qquad [4.14]$$

The equivalent shear stress for a thin hollow tube of wall thickness t is

$$\tau = \frac{2 K_{arb} \delta A}{\pi D^2 t} \qquad [4.15]$$

8. Calculations for the T-bar system are similar but because, in roll, both arms twist the vertical torsion bar in the same direction (unlike the U-bar) it is less stiff. Consequently in *equation [4.10]* the value of L is twice the actual length (instead of half).

EXAMPLE 4.6

A car has a wheel centre rate of 30 N/mm. In order to reduce roll it is required to increase this by 50% with the addition of a U-shaped anti-roll bar. Wheel movement in bump and rebound is 20 mm and it can be assumed that the effective motion ratio of the arb is 1.4.

(a) Determine suitable dimensions of a solid circular arb given a half-length of torsion bar of 125 mm.
(b) Check that the maximum shear stress in the bar does not exceed a limiting value of 900 N/mm^2.

Solution

(a) Wheel centre rate from arb = 30 × 50% = 15 N/mm

$$K_{arb} = 15 \times R_m^2 = 15 \times 1.4^2 = 29.4 \text{ N/mm}$$

Try 12 mm diameter bar - initially ignore bending of arm:

From *equation [4.10]*

$$\text{Stiffness, } K_{bar} = \frac{\pi G D^4}{32 L A^2} = \frac{79\,300 \times \pi \times 12^4}{32 \times 125 \times A^2} = 29.4$$

Hence $A \approx 209 \text{ mm}$

The bending of the arm will reduce the stiffness so try an arm length of 180 mm:

$$K_{bar} = \frac{79\,300 \times \pi \times 12^4}{32 \times 125 \times 180^2} = 39.9 \text{ N/mm}$$

From *equation [4.12]* $$K_{arm} = \frac{3 \times 207\,000 \times \pi \times 12^4}{64 \times 180^3} = 108.4 \text{ N/mm}$$

$$\frac{1}{K_{arb}} = \frac{1}{K_{bar}} + \frac{1}{K_{arm}} = \frac{1}{39.9} + \frac{1}{108.4}$$

$$K_{arb} = 29.2 \text{ N/mm}$$

(b)

Movement of arb, δ = $\dfrac{20}{R_m} = \dfrac{20}{1.4} = 14.3 \text{ mm}$

Shear stress in bar, $\tau = \dfrac{16 k_{arb} \delta A}{\pi D^3} = \dfrac{16 \times 29.2 \times 14.3 \times 180}{\pi \times 12^3}$

$$= 222 \text{ N/mm}^2 \quad < 900 \text{ N/mm}^2$$

Conclusion – a solid circular anti-roll bar of overall length 250 mm, diameter 12 mm and arm length of 180 mm is suitable.

Figure 4.14 shows the printout from a simple spreadsheet that performs the above calculations. The arb stiffness at various arm lengths is tabulated. This spreadsheet is available for download from www.palgrave.com/companion/Seward-Race-Car-Design.

Chapter 4 Springs, dampers and anti-roll

Mod of elasticity, E	207000 N/mm²
Mod of rigidity, G =	79300 N/mm²
Torsion bar length, 2L =	250 mm
Bar diameter, D =	12 mm
Deflection of arb, δ =	14.4 mm

Arm length, A mm	Stiffness N/mm	Stiffness pnds/inch	Stress in rod @ δ mm M/mm²
50	468.71	2676.81	994.63
60	319.57	1825.07	813.78
70	230.59	1316.90	685.05
80	173.44	990.54	588.90
90	134.68	769.14	514.43
100	107.24	612.44	455.13
110	87.15	497.70	406.85
120	72.03	411.35	366.83
130	60.38	344.84	333.15
140	51.24	292.61	304.43
150	43.93	250.91	279.70
160	38.02	217.13	258.18
170	33.17	189.42	239.31
180	29.14	166.43	222.63
190	25.77	147.18	207.81
200	22.92	130.90	194.56
210	20.49	117.03	182.64
220	18.41	105.13	171.88
230	16.61	94.85	162.13
240	15.04	85.92	153.24
250	13.68	78.11	145.12

Figure 4.14
Anti-roll bar calculator

SUMMARY OF KEY POINTS FROM CHAPTER 4

1. Most modern racing cars use spring/dampers hidden inside the bodywork to improve aerodynamics. This generally requires the use of a bell-crank to link the spring/damper to an external pushrod or pullrod. A rocker-arm is an alternative.
2. The use of a soft suspension improves mechanical grip but bottoming-out is a problem as racing cars generally need to operate at the minimum permitted ground clearance. Aerodynamic downforce forces the use of stiffer springs.
3. The motion ratio, R_m, defines the relationship between the movement of the wheel and the movement of the spring. The spring stiffness rate, K_S, is equal to the wheel rate, $K_W \times R_m^2$.
4. Once wheel rates are known the designer can calculate the natural frequencies at which the body (sprung) and the wheels (unsprung) will oscillate. Racing cars tend to have a natural sprung frequency in the range 2 Hz to 5 Hz depending upon the amount of downforce employed.
5. A small amount of spring pre-load can optimise the amount of damper travel available for bump movement.
6. Dampers or 'shocks' prevent excessive oscillations of both the sprung body and the unsprung wheel assemblies after displacement from the neutral position. Critical damping means a return to the neutral position without overshoot. For racing cars the aim is usually to achieve an average of about 0.7 critical

damping at low frequencies. Damping is reduced at higher frequencies to prevent high damping forces from unsettling the car.
7. Mono-tube dampers are considerably more expensive than twin-tube dampers but are better for compact solutions as they have relatively bigger pistons to move more oil over short strokes.
8. Anti-roll systems are used to reduce excessive roll and to provide a means of varying the relative stiffness of one end of the car for tuning purposes. The common form is a U-shaped torsion bar whose stiffness is easily calculated.

5 Tyres and balance

LEARNING OUTCOMES

At the end of this chapter:
- You will be aware of the basic types of racing tyre
- You will understand the important concept of tyre **slip angle** and how this influences understeer and oversteer during cornering
- You will learn the significance of **tyre drag force** and **camber thrust**
- You will be aware of **slip ratio** during acceleration and braking
- You will be able to interpret the results of standard tyre testing
- You will be exposed to a method of presenting tyre behaviour in the form of a mathematical model
- You will be able to calculate individual wheel lateral load transfer during cornering and appreciate how this changes with front and rear suspension roll stiffness
- You will understand the factors that contribute to understeer/oversteer balance and be able to perform the necessary calculations to produce a desired **handling curve**
- You will be able to estimate the actual amount of **jacking** that occurs during cornering

5.1 Introduction

The three basic elements of racing – acceleration, braking and cornering – all require forces to be transmitted from the road to the car via the tyre contact patches. Bastow *et al.* (*ref. 4*) say:

> 'Tyres are the only contact between the car and the ground. Therefore, an understanding of wheel and tyre characteristics is fundamental to the understanding of vehicle behaviour and suspension design.'

Thus the importance of tyres to a competitive car cannot be overestimated. This chapter considers how a car tyre reacts to both lateral forces in cornering and longitudinal forces in acceleration and braking. The chapter continues by showing how selecting appropriate tyres and tuning the stiffness of the front and rear suspensions can produce a car with the desired degree of understeer/oversteer balance.

5.2 Tyres

Most racing cars run on special tyres which cannot be used legally on the public highway. Many formulae adopt **control tyres** which means that all competing cars must use the same make and type. The basic dry-weather tyre is the **slick** which has no tread, to maximise the area of rubber in contact with the road. **Wets** contain significant grooves which are added to enable the tyre to displace water and hence avoid aquaplaning.

The two basic elements of the tyre are the **carcase** and the **wearing course** or tread. The carcase forms the underlying structure of the tyre and consists of cords in a relatively soft rubber matrix. The cords are laid in layers or **plies** and are made of materials such as nylon, steel or Kevlar. In a **radial-ply tyre** the principal cords run at 90° to the direction of travel, however these are supplemented by further reinforcing belts under the tread area. In a **bias-ply** or **cross-ply** tyre the cords run at 45° to the direction of travel. Both forms of carcase construction are still used in racing although the top formulae tend to use radials. The two types produce different driving characteristics with radials providing more ultimate grip but with a sharper decline in grip after the peak. This makes it more difficult to balance a car on radials. Bias-ply tyres are less 'peaky' and consequently more forgiving.

The other element of the tyre, the wearing course, is available in a range of rubber compounds that vary from 'hard' for endurance racing to 'super-soft' for short hill-climbs and sprints. The softer compounds provide more grip but at the expense of reduced life. Tyre companies keep the details of their compounds a closely guarded secret.

We saw in *Figure 1.3* that tyres do not exhibit simple Coulomb friction (i.e. do not have a constant coefficient of friction). The grip from a racing tyre can be considered to have three components:

1. *Friction* – where the friction coefficient varies with both temperature and sliding speed.
2. *Interlocking* – where the rubber deforms around the micro-bumps and depressions in the road surface.
3. *Adhesion* – where a tyre actually sticks itself to the road, particularly when at working temperature (80–110°C).

For components 2 and 3 grip increases with tyre width.

Components 1 and 2 combined with gravity loads from the weight of the car and driver are loosely termed **mechanical grip**. Components 1 and 2 combined with aerodynamic loads from wings etc. form **aero grip**. Component 3 is **chemical grip**.

5.2.1 Cornering – understeer and oversteer

The development of the concept of **slip angle** in the 1930s was responsible for a significant advance in the understanding of how a tyre behaves during

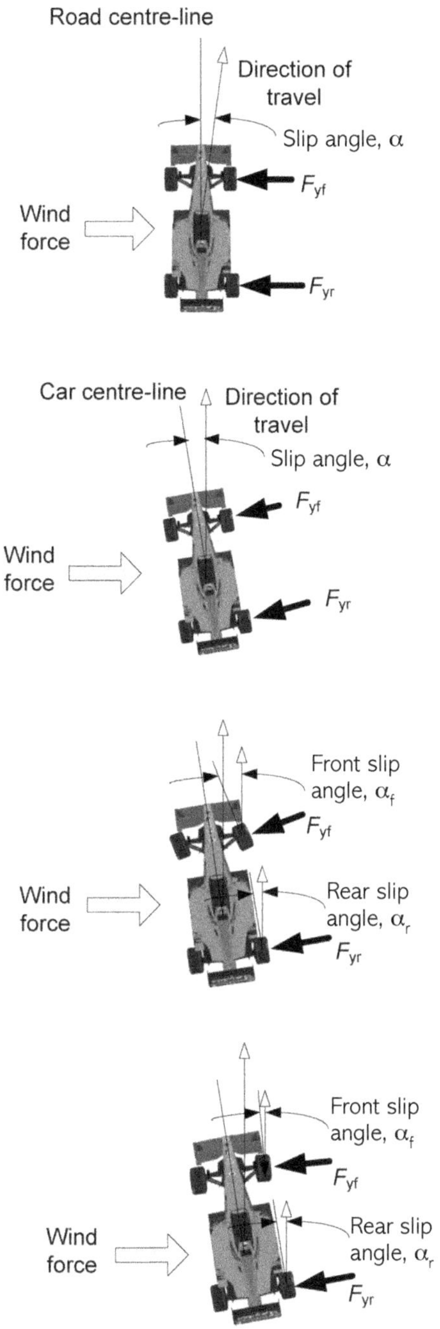

Figure 5.1
Slip angles under lateral wind load

(a) Car pushed off-course by side-wind

(b) Direction corrected – neutral handling car

(c) Direction corrected – understeering car

(d) Direction corrected – oversteering car

cornering, which in turn led to major improvements in vehicle handling. Although this section is concerned with cornering, in order to understand the slip angle concept we will, paradoxically, start by considering a car driving in a straight line, but subjected to a side-wind (this removes the complication of additional front wheel steer-angles).

Figure 5.1a shows a car which was driving in a straight line down the centre of the road until the onset of a significant and steady side-wind. The driver notices that the car is pushed off-course and that it proceeds at some angle to the original direction of motion. The horizontal wind force can be assumed to act at the centre-of-area of the car and is resisted by front and rear grip forces from the tyres, F_{yf} and F_{yr} respectively. This causes distortion of the contact patch which is responsible for the angular deviation. The difference between the direction of motion and the longitudinal axis of the tyre is the **slip angle**, α. The harder the wind blows, the greater the grip forces required and the greater the slip angle developed. The term 'slip angle' is universally adopted but unfortunately it gives the impression of the tyre sliding or skidding across the road. This is not the case. The tyre has not 'let go' and the slip angle is a feature of the twisting and distortion of the rubber in the contact patch (a better term might be 'grip angle'). The slip angle increases with side-force until, for racing tyres, reaching a maximum limit at about 7–10°, at which point the tyre does 'let go' and start to slide across the road surface.

Figure 5.1b shows the situation after some transient steering adjustments by the driver. Under a constant wind force, a steady-state situation has been achieved where the car is proceeding in the right direction but with the whole car oriented at the slip angle, α, relative to the direction of motion. We can conclude that the application of a sideways force to a tyre produces a slip angle and conversely the application of a slip angle produces a sideways force. If, as shown, this steady-state equilibrium is achieved with zero steering angle, i.e the front and rear tyres have the same slip angle, then the car is said to possess **neutral handling**. This implies that the two front tyres develop a combined grip force of F_{yf} and that the rear tyres develop a combined force of F_{yr} when subjected to the same slip angle.

In order to achieve the steady-state condition, the driver may find it necessary to point the front wheels into the wind – *Figure 5.1c*. The front wheel slip angle, α_f, is now greater than the rear wheel slip angle, α_r and the car is said to possess **understeer**. This implies that, given the same front and rear slip angles, the front wheels cannot provide the necessary grip for the wind force. If the driver does not respond by adding steering angle to increase the front slip angle the car deviates from the desired direction.

Conversely the driver may find it necessary to point the front wheels away from the wind – *Figure 5.1d*. The front wheel slip angle, α_f, is now less than the rear wheel slip angle, α_r, and the car is said to possess **oversteer**. Thus in order to achieve the steady-state condition the driver applies a small amount of **negative-lock**. A more formal definition of understeer and oversteer is given on page 149.

We will now turn to cornering. *Figure 5.2a* shows a car moving very slowly around a corner so that the centrifugal force and hence the slip angles are negligible. It can be seen that, in this case, the steering angles are such that the wheels move tangentially to the turning circle. (This is known as Ackermann

Chapter 5 **Tyres and balance**

(a)
Slow moving car with negligible cornering force

(b)
Fast moving car with large cornering force

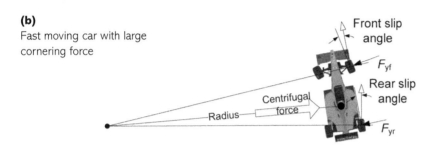

Figure 5.2
Slip angles under cornering

steering and will be covered in more detail in *Chapter 6*.) If the speed of the car is increased, a centrifugal force will develop, the resultant of which will act at the centre-of-mass. This creates equal and opposite grip forces, F_{yf} and F_{yr}, which, in turn, generate slip angles at the tyres. It can be seen that a steady-state condition is reached where the car now proceeds around the corner with its nose pointing inwards towards the centre of the turn. As before, if the front wheel slip angles equal the rear wheel slip angles the car is said to possess **neutral handling**. If the speed of the car is increased beyond the limit, where the centrifugal force equals the maximum grip of the tyres, a neutral handling car should move into a controlled sideways drift. However, as we shall see later, it is often not easy to design a car with consistent neutral handling characteristics under all conditions.

If, in order to achieve this steady-state, the driver needs to increase the front wheel steering angle so that the front slip angles are greater than the rears, the car is said to be **understeering**. If the driver increased the speed of the car without tightening the steering angle the car would proceed on a path with a larger turning radius than that desired. Beyond the limit of grip, even if the driver has applied full steering lock, the car will fail to turn the corner and is likely to leave the track by ploughing straight-on at the outside of a bend. The front tyres have given-way but the rears continue to grip. However when a driver has taken a car beyond the limit, the situation is often recoverable by simply backing off the throttle and slowing down. The reduced centrifugal force means that the front tyres can regain grip.

Conversely, as the speed increases, if the driver needs to back-off the steering angle to achieve a steady-state turn, the car is **oversteering**. Failure

Race car design

to do this results in the car following a tighter radius than desired. If the car exceeds its limit of grip the back-end will give-way and the car will enter a spin – probably leaving the circuit backwards. A skilled driver with fast reactions may be able to rescue the situation by applying opposite lock, however oversteer remains a significantly less stable condition than understeer.

For the above reasons modern production road cars are invariably designed with significant understeer characteristics under all driving conditions. Racing drivers differ in the handling set-up they prefer although they generally like to be close to neutral. For less experienced drivers a modest degree of understeer makes sense and with a rear-wheel-drive car enables the driver to balance the car by accelerating out of corners. We saw in *Chapter 1* that total traction is governed by the traction circle and hence rear lateral grip can be reduced to match the front lateral grip by adding a little forward acceleration.

5.2.2 Tyre cornering forces

Figure 5.3 shows the forces acting on the undriven front right tyre in *Figure 5.2b*. It can be seen that the centrifugal force from the axle passes through the wheel bearings and hence is at the centre of the wheel and perpendicular to the centre-line of the tyre. This is resisted by the grip force from the road contact patch

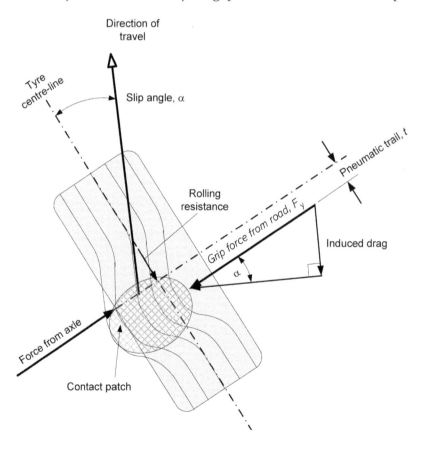

Figure 5.3 Tyre forces under cornering

whose resultant, F_y, is also perpendicular to the tyre centre-line. However it can be seen that, because of distortion in the contact patch, F_y occurs some distance behind the wheel centre-line. This offset is known as **pneumatic trail**, **t**. The value of t reaches a maximum at around half the peak slip angle and then declines as the tyre approaches peak grip. Clearly pneumatic trail causes a moment about the steering axis which tries to straighten the wheel and hence pull the car out of the turn. This moment is referred to as **self-aligning torque**:

$$\text{Self-aligning torque} = F_y \times t \qquad [5.1]$$

The self-aligning torque is transmitted through the steering mechanism and felt by the driver at the steering wheel. By sensing the build-up and then decline of this torque the driver can predict the onset of peak grip. This is what is known as 'good steering feel' and it can be masked on road cars by over-aggressive power-assisted steering.

Also shown in *Figure 5.3* is that the grip force F_y can be resolved into two components – parallel and perpendicular to the direction of motion of the car. The perpendicular component = $F_y \times \cos \alpha$ is essentially the centripetal force and for realistic slip angles, α, can generally be taken as F_y. The parallel component = $F_y \times \sin \alpha$, opposes the direction of motion and is known as **induced tyre drag**. This explains why a driver needs to apply some throttle to maintain a constant speed through a long corner. As mentioned in *Chapter 1*, tyres are also subject to **rolling resistance** which occurs at zero slip angle. Rolling resistance occurs in a free-rolling wheel and is due to energy used in the compression and distortion of the tyre as rubber enters the contact patch. In *Chapter 1* we estimated rolling resistance at about 2% of vertical load for a racing tyre at speed. The total drag on a car is the induced drag for each wheel plus the rolling resistance.

EXAMPLE 5.1

In *Example 1.7* we showed that a F1 car had a total weight of 15 450 N (including downforce) and could develop a total cornering force of 18 540 N on a 100 m radius corner at 49.7 m/s.

(a) Estimate the total tyre drag on the car if all the tyres are assumed to be operating at a slip angle of 8°.
(b) Calculate the engine power required to overcome tyre drag and maintain this constant speed.

(a) Induced drag = $F_{total} \times \sin \alpha$ = 18 540 × sin 8°
     = 2580 N

Rolling resistance = $W_{total} \times 2\%$ = 15450 × 0.02
     = 309 N

$$\text{Total drag} = 2580 + 309 = 2889 \text{ N}$$

(b) $\quad \text{Engine power} = \text{force} \times \text{speed} = 2889 \times 49.7/10^3 \text{ kW}$

$$= \mathbf{143.6 \text{ kW}} \quad (193 \text{ hp})$$

Comment:
This is clearly a significant power requirement and must be added to a possibly larger amount to overcome aerodynamic drag.

5.2.3 Acceleration and braking

Figure 5.4 shows a driven wheel under an acceleration torque. The force from the road pushes the car forward causing longitudinal shear in the contact patch as shown. This causes compression and hence contraction of the tyre tread ahead of the contact patch. This contraction is carried into the contact patch and reduces the effective circumference (and hence radius) of the tyre. Thus for a given vehicle speed a driven wheel must rotate faster than a free-rolling wheel. At the rear of the contact patch the tread recovers by sliding over the road surface and expanding back to normal. The difference in rotational speed of a driven wheel compared to a free-rolling wheel is expressed in the form of a **percentage traction slip ratio**:

$$\text{percentage traction slip ratio} = \left(\frac{\text{speed of driven wheel}}{\text{speed of free-rolling wheel}} - 1 \right) \times 100\%$$

[5.2]

During braking the opposite occurs. *Figure 5.5* shows how the braking torque produces expansion of the tyre tread ahead of the contact patch which causes an effective increase in tyre circumference in the contact patch. Hence a braked wheel rotates more slowly than a free-rolling wheel:

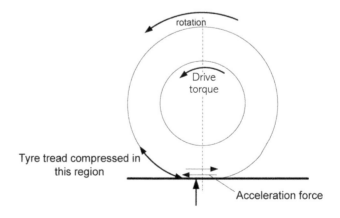

Figure 5.4
Tyre under acceleration

Chapter 5 Tyres and balance

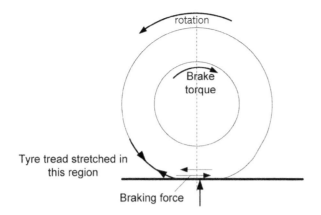

Figure 5.5
Tyre under braking

$$\text{Percentage brake slip ratio} = \left(\frac{\text{speed of braked wheel}}{\text{speed of free-rolling wheel}} - 1\right) \times 100\%$$
[5.3]

If the ratio of longitudinal grip force, F_x, divided by vertical wheel load, F_z, is plotted against the percentage slip ratio (*Figure 5.6*), it can be seen that the peak values of grip occur at about 10–15% slip ratio. This is the figure used when setting-up an automatic **launch control system**. Above this figure the grip level reduces as the wheel moves towards spinning or locking-up. In this case it can be seen that the peak figure of F_x/F_z represents a coefficient of friction of about 1.5. *Figure 5.6* also shows the zones where **traction control** and **anti-lock brake systems** operate, although such systems are not permitted in many formulae. As a wheel moves towards spinning or locking-up it is subject to much larger changes in acceleration than those experienced during normal driving. Wheel speed sensors are used to detect these large accelerations and adjustments made to the brake force or engine power to restore peak grip.

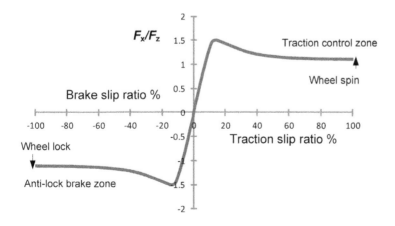

Figure 5.6
Longitudinal grip against slip ratio (%)

127

5.2.4 Camber

Figure 3.3 showed the importance of wheel camber to grip levels and consequently it was argued that designers must maintain careful control of camber during suspension movement. *Figure 5.7* shows a wheel with exaggerated negative camber, $-\gamma$. Left to its own devices a single wheel will rotate about the point, X, which is located at the point where the wheel axis strikes the ground plane. If a cambered wheel is forced to move in the straight-ahead direction the tyre tread will be distorted and a lateral force known as **camber thrust** is generated. For a bias-ply tyre:

$$\text{Camber thrust} \approx F_z \times \gamma \quad \text{(rad)} \qquad [5.4]$$

For a radial tyre the camber thrust is less at say 40% of the above.

If both wheels on an axle have equal negative camber then during straight-line running they will fight each other with opposing camber thrust forces. It would be expected that such tyres would also be producing drag, however the consensus is that this is small.

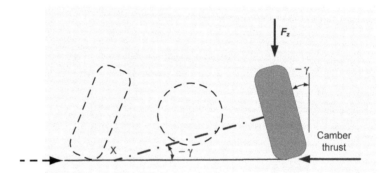

Figure 5.7 Camber and camber thrust

During cornering **lateral load transfer** will cause F_z to increase on the outer wheel and reduce on the inner wheel. Thus the camber thrust force from the outer wheel will dominate and can be combined with the lateral grip force generated by slip angle to combat centrifugal force. However deformation of the tread within the contact patch causes camber thrust to reduce as the slip angle increases. Also suspension movement during chassis roll is likely to reduce negative camber of the outer wheel. Consequently the contribution of camber thrust to total lateral grip is relatively small at the limit.

5.2.5 Tyre testing

Detailed knowledge of tyre data is so important to racing car design and operation that the major teams invest significant sums in commissioning tyre tests from specialist test laboratories. Once paid for, this data is kept highly confidential. FSAE/Formula Student teams are fortunate in being able to join

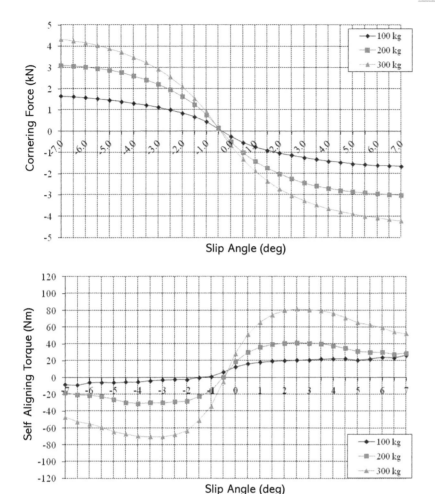

Figure 5.8a
Tyre test data – Rear F3; 180/500R13; 0° camber and 1.7 bar inflation (reproduced with the kind permission of Avon Tyres Motorsport)

a tyre test consortium which for a modest fee will make available data on commonly used tyres.

A tyre-testing machine consists of a rolling road faced with grit-paper to simulate the road surface. Above this the wheel and tyre to be tested are mounted on a fully instrumented and movable axle. The machine enables the contact force between the tyre and the rolling road to be varied. At a typical road speed and tyre pressure, and for several different contact forces, the wheel camber and slip angle are adjusted. The forces and moments generated are logged electronically.

One of the few tyre companies to make test data available is Avon (*ref. 3*) and *Figures 5.8a* and *5.8b* show some of their data in its commonest form – graphs showing the variation of cornering force (lateral grip) and self-aligning torque with slip angle. The individual curves represent different increments of vertical tyre load. These curves relate to the rear 2001 control tyre for the British F3 Championship. There is a set of graphs for each of several camber angles but only those for 0° camber are shown in this instance. Points to note are:

Figure 5.8b
Tyre test data – Rear F3; 250/570R13; 0° camber and 1.7 bar inflation. Reproduced with the kind permission of Avon Tyres Motorsport

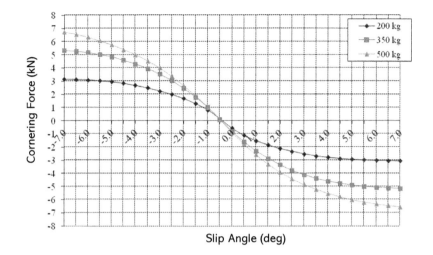

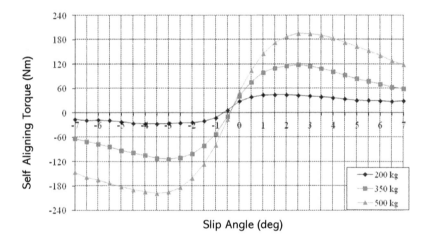

- Although the vertical load on the tyre is increased in equal increments, it can be seen that the resulting cornering force curves become progressively closer together as the load increases. This is as a result of tyre sensitivity as described in *Chapter 1*.
- The tests extend to slip angles of ±7° and it is a pity that the curves do not extend beyond the point of peak grip which generally occurs at about 10° to 11° with these tyres. This would provide more confidence when calculating car balance.

We will show how these curves are used to determine the handling balance of a car later in this chapter.

5.2.6 Tyre modelling

An alternative method of presenting tyre test data is in the form of a mathematical tyre model which is essentially a formula or equation. The principal advantage of a mathematical model over the previous graphical representation is that it is easy to automate calculations within a computer program. There are several competing tyre models but we will consider the most widely used which is the **Pacejka '96 model**, often referred to as the **magic formula** (*ref. 17*). The 'magic' refers to the fact that essentially the same formula can be used to predict both lateral and longitudinal grip as well as self-aligning torque and combinations of lateral and longitudinal grip. It is known as a 'semi-empirical' formula which means that it is partly based on theoretical considerations but is populated by coefficients derived from real experimental tests. The technique is based upon an equation that defines the shape of the relevant curve at a particular value of vertical load known as the **nominal wheel load**, F_{Z0}. Various coefficients are then used to modify the shape of the curve for different wheel loads and camber values. (*Appendix 1* considers Pacejka tyre coefficients in more detail.) We will only consider its application in relation to pure lateral grip where the basic form which outputs the lateral grip force, F_y, is:

$$F_y = D_y \sin[C_y \tan^{-1}\{B_y\alpha_y - E_y(B_y\alpha_y - \tan^{-1}(B_y\alpha_y))\}] + S_{Vy} \quad [5.5]$$

In the above:

D_y is the **peak value** and is related to the friction coefficient:

$$D_y = F_Z(p_{DY1} + p_{DY2}df_Z)(1 - p_{DY3}\gamma_y^2)\lambda_{\mu y} \quad [5.6]$$

where F_Z = actual vertical wheel load (input in N)
$p_{CY1}...p_{VY4}$ = Pacejka coefficients obtained from tyre test data
(18 such coefficients are required for lateral grip – *Table 5.1*)

$$df_Z = \text{normalised change in vertical load} = \frac{F_Z - F_{Z0}}{F_{Z0}} \quad [5.7]$$

γ_y = actual wheel camber (input in radians)

$\lambda_{\mu y}$ = user scaling factor on the friction coefficient.

(Several such 'λ' scaling factors are available to enable the user to adapt the formula for track conditions that differ from the test conditions. Generally the value is set to 1, however the above $\lambda_{\mu y}$ factor is particularly useful if the track surface is not as 'grippy' as the grit-paper used on the test rolling road. This is generally relevant to the Formula SAE tyre consortium data which over-estimates the friction coefficient. A $\lambda_{\mu y}$ value of about 0.6–0.7 appears to give more realistic results.)

C_y is the **shape factor** which together with the **curvature factor**, E_y, determines the appearance of the curve in the region of the peak. (In this context 'shape' refers to the shape of the curve, not the tyre.)

$$C_y = p_{CY1}\lambda_{Cy} \qquad [5.8]$$

B_y is the **stiffness factor** and together with C_y and D_y determines the slope of the curve near the origin.

$$B_y = \frac{P_{Ky1}F_{Z0}\sin\left[2\tan^{-1}\left\{\dfrac{F_Z}{P_{Ky2}F_{Z0}\lambda_{Z0}}\right\}\right](1 - P_{Ky3}|\gamma|)\lambda_{FZ0}\lambda_{Kya}}{C_y D_y} \qquad [5.9]$$

and $\qquad \alpha_y = \alpha + S_{Hy} \qquad [5.10]$

α = actual slip angle (input in radians)

S_{Hy} is a term quantifying the horizontal shift of the curve at the origin:

$$S_{Hy} = (p_{HY1} + p_{HY2}df_Z + p_{HY3}\gamma_y)\lambda_{Hy} \qquad [5.11]$$

Table 5.1 Pacejka coefficients – Avon British F3 Tyres

	Front	Rear	Description
F_{Z0} =	2444	3850	Nominal load (N)
p_{CY1} =	0.324013	0.558238	Shape factor
p_{DY1} =	–3.674945	–2.23053	Lateral friction, μ_y
p_{DY2} =	0.285134	0.090785	Variation of friction with load
p_{DY3} =	–2.494252	–5.71836	Variation of friction with camber squared
p_{EY1} =	–0.078785	–0.40009	Lateral curvature at F_{Z0}
p_{EY2} =	0.245086	0.569694	Variation of curvature with load
p_{EY3} =	–0.382274	–0.26276	Zero order camber dependency of curvature
p_{EY4} =	–6.25570332	–29.3487	Variation of curvature with camber
p_{KY1} =	–41.7228113	–28.2448	Maximum value of stiffness K_y/F_{Z0}
p_{KY2} =	2.11293838	1.331304	Normalised load at which K_y reaches max. value
p_{KY3} =	0.150080764	0.255683	Variation of K_y/F_{Z0} with camber
p_{HY1} =	0.00711	0.00847	Horizontal shift S_{Hy} at F_{Z0}
p_{HY2} =	–0.000509	0.000594	Variation of S_{Hy} with load
p_{HY3} =	0.049069131	0.042	Variation of S_{Hy} with camber
p_{VY1} =	–0.00734	0.0262	Vertical shift S_{Vy} at F_{Z0}
p_{VY2} =	–0.0778	–0.0791	Variation of S_{Vy} with load
p_{VY3} =	–0.0641	–0.08552	Variation of S_{Vy} with camber
p_{VY4} =	–0.6978041	–0.44481	Variation of S_{Vy} with camber and load

E_y is the **curvature factor**:

$$E_y = (p_{EY1} + p_{EY2}df_Z)\{1 - (p_{EY3} + p_{EY4}\gamma_y)\text{sgn}(\alpha_y)\}\lambda_{Ey} \qquad [5.12]$$

where sgn(α_y) takes the value of 1 if α_y is positive and –1 if α_y is negative. S_{Vy} is a term quantifying the vertical shift of the curve at the origin:

$$S_{Vy} = F_Z\{p_{VY1} + p_{VY2}df_Z + (p_{VY3} + p_{VY4}\,df_Z)\gamma_y\}\lambda_{Vy}\lambda_{Kya} \qquad [5.13]$$

Clearly the repeated application of this complex formula by hand would be tedious, however it is readily implemented in a computer program. A simple spreadsheet which implements the above equations is available for download from www.palgrave.com/companion/Seward-Race-Car-Design.

Table 5.1 gives the nominal load plus the 18 Pacejka coefficients for the front and rear Avon F3 tyres shown graphically in *Figures 5.8a* and *b*. The excessive accuracy of the coefficient values indicates that they were derived from a specialised computer program that operates on the raw test data. Inspection of such data often reveals little apparent correlation between the numerical values and the stated description of particular coefficients; however they still usually work satisfactorily. A demonstration of the use of the Pacejka formula is provided in *Example 5.3*.

5.3 Balancing a racing car

5.3.1 Individual wheel loads during lateral load transfer

In *Chapter 1* we saw how, during cornering, a centrifugal force acting through the centre of mass causes load to transfer from the inside to the outside wheels. We learned how to calculate a total value for this transfer. We also saw that, because of tyre sensitivity, load transfer across an individual axle causes a reduction in lateral grip at that axle. Consequently being able to vary the proportion of load transferred at the front and rear of a car is an important means of tuning the oversteer/understeer balance of a car. We therefore need to be able to calculate individual wheel loads during cornering.

The first step is to split the mass of the car into three elements: the sprung mass, including the driver, and the front and rear unsprung masses. *Figure 5.9* shows the geometrical information required. As shown, the location of the centre of the unsprung masses is usually considered to be on the centre-line of the car at the wheel centres. The **roll axis** is the line joining the roll-centres at each end of the car which are at heights of h_{rcf} and h_{rcr} above ground. The distance, h_a, from the sprung mass, m_s, to the roll axis is important for determining the **roll couple**. This should be the perpendicular distance to the roll axis, however, because the inclination of the roll axis is generally small, it is normally taken to be the vertical distance.

Race car design

Figure 5.9
Lateral load transfer geometry

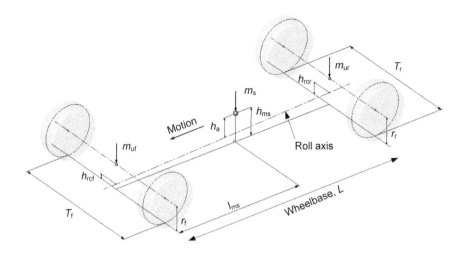

Step 1 – Static wheel loads

Static wheel loads can be calculated either from the combined weight of the vehicle and the position of its centre of mass as shown in *Example 1.2* or from the sprung and unsprung components as follows:

ΣM *about front axle* $\quad W_r \times L = 0.5g(m_{ur}L + m_s l_{ms})$

$\qquad$ Rear wheel loads, $\mathbf{W_r = 0.5g(m_{ur} + m_s l_{ms}/L)}$ $\qquad$ [5.14]

$\Sigma V = 0 \quad$ Front wheel loads, $\mathbf{W_f = 0.5g(m_{uf} + m_{ur} + m_s) - W_r}$ $\qquad$ [5.15]

For a car with aerodynamic downforce the vertical wheel loads are increased but the magnitude of downforce increases with the square of the vehicle speed. Also the distribution between the front and rear axles depends upon both the location of the centre of aerodynamic vertical force and the height of the resulting aerodynamic drag force which causes front/rear load transfer. The resulting distribution should generally be close to that which results from the vehicle mass so that the balance of the car does not change significantly with speed. It is necessary to check the balance of the car both with zero downforce (slow hairpin) and at the speed of the fastest corner.

Step 2 – Unsprung mass lateral force

Lateral load transfer calculations are normally carried out in increments of lateral *g*. At say 1.5 lateral *g*, all the vehicle masses (kg) are multiplied by a lateral acceleration, A_y, of 1.5×9.81 to get lateral forces (N).

From *Figure 5.10* for the front axle:

Take moments about the inner wheel contact patch:

$$A_y m_{uf} r_f = \Delta W_{uf} T_f$$

Chapter 5 Tyres and balance

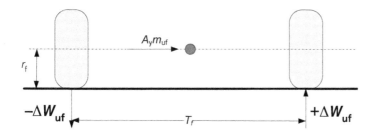

Figure 5.10
Lateral load transfer from front unsprung mass

Load transfer

$$\Delta W_{uf} = \frac{A_y m_{uf} r_f}{T_f} \qquad [5.16]$$

Similarly for the rear axle:

Load transfer

$$\Delta W_{ur} = \frac{A_y m_{ur} r_r}{T_r} \qquad [5.17]$$

Step 3 – Sprung mass lateral force through the suspension links

The left of *Figure 5.11* shows the lateral force from the sprung mass, $A_y m_s$, acting through its centroid which is distance h_a above the roll axis. This is directly equivalent to the same force being applied at the roll axis plus a roll couple, $C = A_y m_s h_a$ as shown on the right in *Figure 5.11*. Because the force acts at the roll axis it does not cause any roll movement in the springs. It is transmitted down to the ground via forces in the suspension links. This is the focus of *step 3*. The roll couple is dealt with in *step 4*.

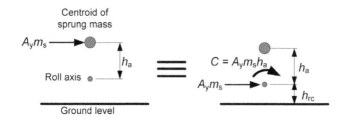

Figure 5.11
Lateral load transfer from the sprung mass

The lateral force from the sprung mass is simply divided between the front and rear axles depending upon the position of the sprung mass centroid, l_{ms}.

Sprung mass force resisted at the front axle = $A_y m_s \times (L - l_{ms})/L$

Resulting load transfer, $\Delta W_{sff} = (A_y m_s \times (L - l_{ms})/L) \times (h_{rcf}/T_f) \qquad [5.18]$

Similarly for the rear axle:

Load transfer, $\Delta W_{sfr} = (A_y m_s \times l_{ms}/L) \times (h_{rcr}/T_r) \qquad [5.19]$

135

Step 4 – Sprung mass roll couple through the springs

Height of roll axis at sprung mass centroid $= h_{rcf} + l_{ms}(h_{rcr} - h_{rcf})/L$

$$\therefore h_a = h_{ms} - h_{rcf} - l_{ms}(h_{rcr} - h_{rcf})/L$$

$$\text{Roll couple, } C = A_y m_s h_a$$

The roll couple is distributed between the front and rear axles in proportion to the roll rates of the axles. We know from *equation [4.5]* that:

$$\text{Roll rate, } K_\phi = \frac{T^2 K_R}{114.6 \times 10^3} \text{ Nm/deg}$$

where T is the relevant wheel track dimension (mm) and K_R is the appropriate wheel ride rate (Nmm).

The wheel ride rate takes into account the stiffness of suspension springs, any anti-roll bars and tyres. As we shall see, it is by changing the ratio of wheel ride rates at each end of the car that the handling balance of a car is tuned.

$$\text{Roll couple resisted at the front axle} = \frac{K_{\phi f}}{K_{\phi f} + K_{\phi r}} \times C$$

$$\text{Resulting load transfer, } \Delta W_{scf} = \left(\frac{K_{\phi f}}{K_{\phi f} + K_{\phi r}} \times C \right) / T_f \qquad [5.20]$$

Similarly for the rear axle:

$$\text{Load transfer, } \Delta W_{scr} = \left(\frac{K_{\phi r}}{K_{\phi f} + K_{\phi r}} \times C \right) / T_r \qquad [5.21]$$

Step 5 – Total lateral load transfer

The wheel loads from *steps 1 to 4* are now combined to give the total lateral load transfer:

$$\text{Front load transfer, } \Delta W_f = \Delta W_{uf} + \Delta W_{sff} + \Delta W_{scf}$$

$$\text{Rear load transfer, } \Delta W_r = \Delta W_{ur} + \Delta W_{sfr} + \Delta W_{scr}$$

$$\text{Front inner wheel load, } W_{fi} = W_f - \Delta W_f$$

$$\text{Front outer wheel load, } W_{fo} = W_f + \Delta W_f$$

$$\text{Rear inner wheel load, } W_{ri} = W_r - \Delta W_r$$

$$\text{Rear outer wheel load, } W_{ro} = W_r + \Delta W_r$$

It can be seen that the individual load transfer terms are all proportional to the lateral acceleration, A_y. It is therefore an easy matter to modify the total load transfer figures for different lateral g increments.

EXAMPLE 5.2

For a car with the following data, determine the wheel loads when cornering at 1.25 lateral g:

	Front	Rear
Wheel radius, r (mm)	270	280
Wheel track, T (mm)	1550	1500
Height roll centre, h_{rc} (mm)	66	77
Ride rate, K_R (N/mm)	34.6	33.5
Unsprung mass, M_u (kg)	32.4	48.0
Wheelbase, L (mm)	2290	
Sprung mass, M_s (kg)	319.6	
Height sprung mass, h_{ms} (mm)	301	
Dist. from fr. axle to M_s, l_{ms} (mm)	1343	

Step 1 – Static wheel loads

Rear wheel loads, $R_r = 0.5g(m_{ur} + m_s l_{ms}/L)$

$\quad = 0.5 \times 9.81 \times (48.0 + 319.6 \times 1343/2290)$

$\quad = $ **1155 N**

Front wheel loads, $R_f = 0.5g(m_{uf} + m_{ur} + m_s) - R_r$

$\quad = 0.5 \times 9.81 \times (32.4 + 48.0 + 319.6) - 1155$

$\quad = $ **807 N**

Step 2 – Unsprung mass lateral force

$\Delta W_{uf} = \dfrac{A_y m_{uf} r_f}{T_f} \quad = 1.25 \times 9.81 \times 32.4 \times 270 / 1550 \; = \; $ **69 N**

$\Delta W_{ur} = \dfrac{A_y m_{ur} r_r}{T_r} \quad = 1.25 \times 9.81 \times 48.0 \times 280 / 1500 \; = \; $ **110 N**

Step 3 – Sprung mass lateral force through the suspension links

Load transfer, $\Delta W_{sff} = (A_y m_s \times (L - l_{ms})/L) \times (h_{rcf}/T_f)$

$\quad = (1.25 \times 9.81 \times 319.6 \times (2290 - 1343)/2290) \times 66/1550$

$\quad = $ **69 N**

Similarly for the rear axle:

Load transfer, $\Delta W_{sfr} = (A_y m_s \times l_{ms}/L) \times (h_{rcr}/T_r)$

$= (1.25 \times 9.81 \times 319.6 \times 1343/2290) \times 77/1500$

= **118 N**

Step 4 – Sprung mass roll couple through the springs

$$h_a = h_{ms} - h_{rcf} + l_{ms}(h_{rcr} - h_{rcf})/L$$

$$= 301 - 66 - 1343 \times (77 - 66)/2290$$

$$= 228.5 \text{ mm}$$

Roll couple, $M_R = A_y m_s h_a$ $= 1.25 \times 9.81 \times 319.6 \times 228.5$

$= 895\,513$ Nmm

Roll rate, $K_{\phi f} = \dfrac{T_f^2 K_{Rf}}{114.6 \times 10^3}$ Nm/deg $= 1550^2 \times 34.6/(114.6 \times 10^3)$

$= 725$ Nm/deg

Roll rate, $K_{\phi r} = \dfrac{T_r^2 K_{Rr}}{114.6 \times 10^3}$ Nm/deg $= 1500^2 \times 33.5/(114.6 \times 10^3)$

$= 658$ Nm/deg

i.e. for this car $725/(725 + 658) \times 100 = 52.4\%$ of the roll couple will be transferred at the front wheels.

Load transfer, $\Delta W_{scf} = \left(\dfrac{K_{\phi f}}{K_{\phi f} + K_{\phi r}} \times M_R \right) / T_f$

$= \left(\dfrac{725}{725 + 658} \times 895\,513 \right) / 1550$

= **303 N**

Load transfer, $\Delta W_{scf} = \left(\dfrac{K_{\phi r}}{K_{\phi f} + K_{\phi r}} \times M_R \right) / T_r$

$= \left(\dfrac{658}{725 + 658} \times 895\,513 \right) / 1550$

= **284 N**

Step 5 – Total lateral load transfer

Front load transfer, $\Delta_f = \Delta W_{uf} + \Delta W_{sff} + \Delta W_{scf}$

$= 69 + 69 + 303$ $= $ **441 N**

$$\text{Rear load transfer, } \Delta_r = \Delta W_{ur} + \Delta W_{sfr} + \Delta W_{scr}$$
$$= 110 + 118 + 284 \qquad = 512 \text{ N}$$

Front inner wheel load, $W_{fi} = W_f - \Delta W_f \quad = 807 - 441 \quad = \mathbf{366\ N}$

Front outer wheel load, $W_{fo} = W_f + \Delta W_f \quad = 807 + 441 \quad = \mathbf{1248\ N}$

Rear inner wheel load, $W_{ri} = W_r - \Delta W_r \quad = 1155 - 512 \quad = \mathbf{643\ N}$

Rear outer wheel load, $W_{ro} = W_r + \Delta W_r \quad = 1155 + 512 \quad = \mathbf{1667\ N}$

A simple spreadsheet which implements the above calculations is available for download from www.palgrave.com/companion/Seward-Race-Car-Design.

5.3.2 Factors affecting the understeer/oversteer balance of a car

The forces that act on a car during cornering can either be resolved perpendicular to the direction of motion (*Figure 5.3*) or perpendicular to the axis of the car. Both give the same result but the latter, as shown in *Figure 5.12*, makes it easier to understand the influence of induced tyre drag on car balance. Also, for

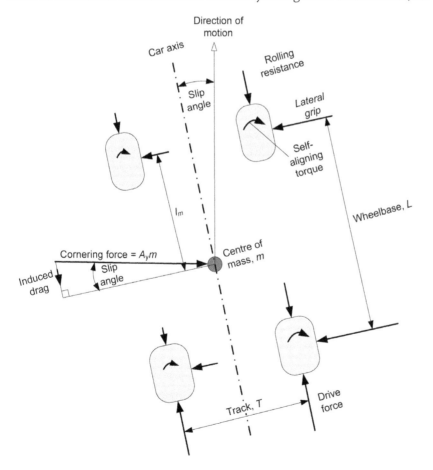

Figure 5.12
Factors affecting understeer/oversteer balance – **neutral handling car**

clarity, the car is shown on a fast large-radius corner where steering angles can be ignored. At each wheel there is a lateral grip force, an associated self-aligning torque and a rolling resistance force. In addition the driven wheels have a drive force which is necessary to maintain a constant velocity in the presence of induced drag and rolling resistance. The procedure to evaluate the understeer/oversteer balance is carried out in increments of lateral *g* as follows:

1. Determine the basic lateral grip forces required at the front and rear axles to resist the cornering force.
2. Modify these values to take account of the other forces and torques that affect balance. In this case any force or torque that produces a clockwise moment about the centre of mass will try to pull the car out of the bend and hence contribute to understeer. Vice versa for oversteer.
3. Knowing individual wheel loads, from the previous section, use tyre data to determine the resulting tyre slip angles that are generated at the front and rear axles.
4. Assess front and rear slip angles to determine degree of understeer/oversteer and if necessary adjust relative front and rear roll rates and repeat.

The forces shown in *Figure 5.12* will now be considered in more detail.

Cornering force

This is the centrifugal force acting at the centre of mass:

$$\text{Cornering force} = A_y m$$

It will act perpendicular to the direction of motion of the vehicle which, for a neutral car, is at the slip angle, α, to the car axis. It is shown resolved perpendicular and parallel to the vehicle axis. The parallel component, $A_y m \sin \alpha$, is equal to the total induced tyre drag. The perpendicular component, $A_y m \cos \alpha$, can be assumed to be equal simply to $A_y m$ for realistic values of α.

ΣM *about front axle*   Rear lateral grip required, $F_{yr} = A_y m l_m / L$ [5.22]

Front lateral grip required, $F_{yf} = A_y m - F_{yr}$ [5.23]

Rolling resistance

From *section 5.2.2* rolling resistance can be estimated at 2% of the vertical load on the tyre. This means that there will be more rolling resistance on the heavily loaded outer wheels. This produces a clockwise moment about the centre of mass and hence contributes to *understeer*. The difference between the rolling resistance forces on each side of the car is thus equal to about 2% of the difference between wheel loads at each side of the car. To compensate for this, front grip F_f will increase and rear grip F_r will reduce.

Understeering moment, M_u

$$= 0.02 \times [((W_{fo} - W_{fi}) \times T_f) + ((W_{ro} - W_{ri}) \times T_r)] \quad [5.24]$$

Change in grip, $\Delta F_f = -\Delta F_r \qquad = M_u/L \qquad [5.25]$

Self-aligning torque

It can be seen from *Figure 5.12* that all the self-aligning torques act in a clockwise direction also contributing to understeer. (Remember this is because pneumatic trail causes the resultant lateral grip force to act behind the wheel centre.) A problem is that the magnitude of self-aligning torques depends upon slip angles – see *Figure 5.8* – which are not yet known. It is therefore usually necessary to estimate front and rear slip angles in order to obtain initial values for self-aligning torques.

Understeering moment, $M_u = \Sigma$ self-aligning torques

Change in grip, $\Delta F_f = -\Delta F_r = M_u/L$

Induced tyre drag

It can be seen from *Figure 5.12* that, for a *neutral* handling car, the total induced drag can be considered to act at the centre of mass and hence does not contribute to understeer or oversteer. *Figure 5.13* shows the situation with an *understeering* car. In order to balance the car the driver has had to turn the steering wheel so that the front slip angle is greater than the rear (this is *not* a steering angle). Now if, as before, the forces are resolved parallel and perpendicular to the car axis, it can be seen that the front tyres have a drag component which is greater on the heavily loaded outside wheel. This pulls the car out of the corner and hence contributes to further understeer. It can be seen that the angle between the front wheel lateral grip force and the perpendicular to the car axis is the difference between the front and rear slip angles. The difference between inner and outer front wheel lateral grip forces can be approximated to the difference between the front wheel loads ($W_{fo} - W_{fi}$), multiplied by the number of lateral g forces.

Understeering moment, M_u

$$= (W_{fo} - W_{fi}) \times A_y/g \times \sin(\alpha_f - \alpha_r)T_f \qquad [5.26]$$

Change in grip, $\Delta F_f = -\Delta F_r \qquad = M_u/L \qquad [5.27]$

For an *oversteering* car the signs are opposite and the forces contribute to further oversteer.

Race car design

Figure 5.13
Factors affecting understeer/oversteer balance –
understeering car

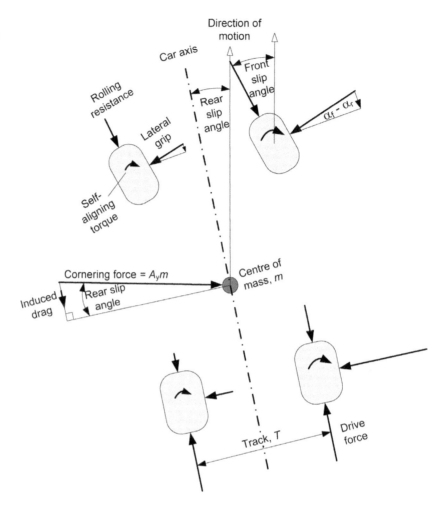

Drive force

There are two ways that drive force can affect balance. The first is that the forces applied to the road by the two driven wheels may be different depending upon the type of **differential** used. A conventional **'open' differential** equalises both torque outputs and so has no significant effect on car balance. **Limited slip** and **torque biasing** differentials tend to bias torque towards the slowest wheel which is, of course, the inside wheel. This contributes to *understeer*, but would have a negligible effect on fast corners where there is little speed difference between the wheels. This effect will be ignored in this analysis.

The second influence that drive force has on car balance is related to the traction circle concept covered in *section 1.8*. We saw that ultimate lateral grip is reduced when cornering is combined with acceleration or braking. Although in this case we are only talking about maintaining a constant speed, this still requires some drive traction to overcome tyre drag and this

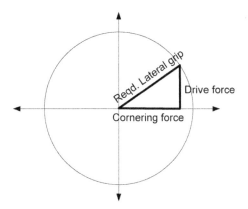

Figure 5.14
Compensating for drive force

is enough to reduce the grip available for cornering at the driven wheels. The effect becomes significant at slip angles above 2°. The presence of a drive force increases the slip angle compared to an axle with lateral force only. For a rear-wheeled-drive car this contributes to *oversteer*. A convenient way of compensating for this is to increase the total grip required at the driven axle. *Figure 5.14* shows a suggested method for doing this based on the traction circle where the required total grip is the hypotenuse of the triangle.

Figure 5.13 Drive force = induced drag + rolling resistance

$$= A_y m \sin \alpha_r + 2W_f \sin(\alpha_f - \alpha_r) + 0.02mg \quad [5.28]$$

For a rear-wheeled-drive car:

$$\text{Reqd. rear lat. grip} = \sqrt{((\text{drive force})^2 + F_{yr}^2)} \quad [5.29]$$

The effect of the drive force required to overcome aerodynamic drag is more complex. Firstly, the magnitude of aerodynamic drag is speed dependent. Secondly, because the drag force is often high-up, as a result of the rear wing, load transfer from the front to the rear of the car increases rear grip (*understeer effect*) but the resulting required drive force weakens rear grip (*oversteer effect*). In practice it seems desirable to balance the aerodynamic loads so that the above two effects cancel out.

Although not part of steady-state cornering, which is the focus of this section, this is an appropriate point to mention two other related 'drive force' phenomena:

- Firstly, because the application of drive force produces an oversteer effect, in a powerful rear-drive car a skilled driver can change an understeering car into a neutral or even oversteering one by accelerating through corners to weaken rear grip. This is one reason why many keen drivers prefer rear-drive cars. Unskilled drivers apply too much throttle and spin!
- Secondly, hard acceleration from a slow hair-pin into a fast corner produces two opposing effects. Rear load transfer strengthens rear grip by effectively creating a larger traction circle whereas the required drive force weakens

Race car design

rear grip. If, when at the traction limit, the driver suddenly backs-off the throttle mid-corner, the reduction in rear grip from loss of load transfer can be greater than the gain in grip from reduced drive force. The result is a spin from **power-off oversteer**.

Lift from steering geometry

Chapter 6 indicates that during cornering **caster** and **kingpin inclination** cause the lightly loaded inner wheel to drop as the steering wheel is turned. This has the effect of lifting that corner of the car and hence changing the distribution of wheel loads. Lateral load transfer is reduced at the front and increased at the rear – hence contributing to *oversteer*. The effect will be more significant with stiffer suspensions and tighter turns but with a typical 50 m radius turn requiring about 2.5° of steering angle the jacking difference across the front of the car is about 2 mm. With ride rates of 30 N/mm this generates about 15 N of lateral load transfer at each end of the car. The calculated wheel loads can be adjusted by the relevant amount.

EXAMPLE 5.3

The car in *Example 5.2* had the following wheel loads in N when cornering at 1.25 lateral g:

Front inside	366
Front outside	1248
Rear inside	643
Rear outside	1667

Wheelbase = 1990 mm, front track = 1550 mm, rear track = 1500 mm and total mass = 400 kg.

If the Avon F3 tyres shown in *Figures 5.8a* and *b* are used, check the understeer/oversteer balance of the car:

(a) using the graphical test data in *Figures 5.8a* and *b*,
(b) using the Pacejka model data in *Table 5.1* and the available software program.

At 1.25 lateral *g* the basic lateral grip forces required at each end of the car are 1.25 × total axle loads:

Front basic lateral grip force = 1.25 × (366 + 1248) = **2018 N**

Rear basic lateral grip force = 1.25 × (643 + 1667) = **2888 N**

Chapter 5 Tyres and balance

Rolling resistance

Understeering moment, M_u

$$= 0.02 \times [((W_{fo} - W_{fi}) \times T_f) + ((W_{ro} - W_{ri}) \times T_r)]$$
$$= 0.02 \times [((1248 - 366) \times 1550) + ((1667 - 643) \times 1500)]$$
$$= 58\,060 \text{ Nmm}$$

Change in grip, $\Delta F_f = -\Delta F_r$ $\qquad = M_u/L$
$$= 58\,060/2990 \qquad = \pm 19.4 \text{ N}$$

Self-aligning torque

We will initially estimate the front slip angle at 3° and the rear at 2°.

From *Figure 5.8a* Front inner torque ≈ 7 Nm
Front outer torque ≈ 25 Nm
From *Figure 5.8b* Rear inner torque ≈ 13 Nm
Rear outer torque ≈ 33 Nm
Total, M_u ≈ 78 Nm $\qquad = 78\,000$ Nmm

Change in grip, $\Delta F_f = -\Delta F_r$ $\qquad = M_u/L$
$$= \pm 78\,000/2990 \quad = \pm 26.1 \text{ N}$$

Induced tyre drag

Understeering moment, M_u

$$= (W_{fo} - W_{fi}) \times A_y/g \times \sin(\alpha_f - \alpha_r)\, T_f$$
$$= (1248 - 366) \times 1.25 \times \sin(3° - 2°) \times 1550$$
$$= 29\,820 \text{ Nmm}$$

Change in grip, $\Delta F_f = -\Delta F_r$ $\qquad = M_u/L$
$$= \pm 29\,820/2990 \qquad = \pm 10.0 \text{ N}$$

Drive force

Drive force $= A_y m \sin \alpha_r + 2W_f \sin(\alpha_f - \alpha_r) + 0.02mg$
$$= 1.25 \times 9.81 \times 400 \sin 2° + 2 \times 807 \sin(3° - 1°) + 0.02 \times 400 \times 9.81$$
$$= 171 + 28 + 78 \qquad\qquad = 277 \text{ N}$$

Reqd. rear lat. grip $= \sqrt{((\text{drive force})^2 + F_{yr}^2)}$
$$= \sqrt{(277^2 + 2888^2)} \qquad\qquad = \mathbf{2901 \text{ N}}$$

Race car design

Summary

Required front lateral grip = 2018 + 19.4 + 26.1 + 10.0   **= 2073 N**

Required rear lateral grip = 2901 − 19.4 − 26.1 − 10.0   **= 2846 N**

Jacking from steering geometry

As suggested above, reduce front load transfer by 15 N per wheel and increase the rear. Revised wheel loads become:

Front inside, W_{fi} 381

Front outside, W_{fo} 1233

Rear inside, W_{ri} 628

Rear outside, W_{ro} 1682

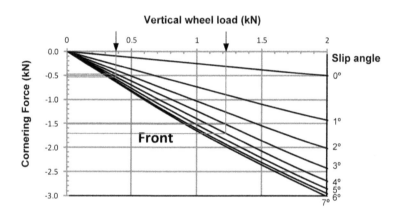

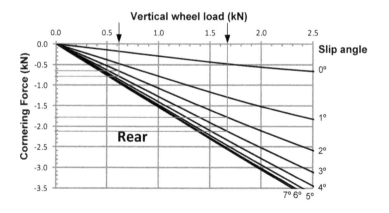

Figure 5.15
Tyre test data. Redrawn from *Figures 5.8a and b*

Chapter 5 Tyres and balance

Checking the understeer/oversteer balance

(a) Using test charts

The following check is possible using the original tyre data curves given in *Figures 5.8a* and *b*; however interpolation is easier if, as suggested by Daniels (*ref. 5*), the charts are redrawn with different axes. *Figure 5.15* shows such charts – the test data is the same but the axes are now 'vertical wheel load' and 'cornering force' and the range of the graphs is restricted to the relevant sector. Individual curves represent slip angles.

The above wheel loads are indicated by the arrows on the horizontal axes. The cornering forces are first read off the charts for the assumed slip angles of 3° front and 2° rear:

Front cornering force (3°) = 480 + 1510 = 1990 < 2073

Rear cornering force (2°) = 670 + 1800 = 2470 < 2846

It can be seen that both slip angle estimates are a little low. If we add 1° to both:

Front cornering force (4°) = 520 + 1700 = 2220 > 2073

Rear cornering force (3°) = 800 + 2120 = 2920 > 2846

By linear interpolation:

$$\text{Front slip angle} = 3 + (2073 - 1990)/(2220 - 1990)$$
$$= \mathbf{3.4°}$$
$$\text{Rear slip angle} = 2 + (2846 - 2470)/(2920 - 2470)$$
$$= \mathbf{2.8°}$$

At this point it should be considered whether or not it is necessary, in the light of the errors in the initial slip angle estimates, to go back and correct the various adjustments to the basic lateral grip forces and repeat the analysis. In this case it would make little difference. In addition we should really have evaluated the individual wheel cambers at 1.25 lateral *g* and used appropriate tyre data curves.

Conclusion At 1.25 lateral *g* the car exhibits (3.4° − 2.8°) = 0.6° of mild understeer.

(b) Using Pacejka tyre model

Table 5.2 shows the Pacejka model spreadsheet, A data column has been created for each wheel which is populated with the appropriate parameters from *Table 5.1* at the top. The user simply changes the slip angle values inside the dashed box until the sums of the cornering forces at each axle equal the target values – i.e. 2073 front and 2846 rear. It can be seen that the resulting slip angles are somewhat lower than those obtained from the graphs – i.e.

Race car design

Table 5.2 Pacejka spreadsheet – from Avon Tyres Motorsport

Lateral Force	Front 180/550	Front 180/550	Rear 250/570	Rear 250/570
PCy1	0.324013	0.324013	0.558238	0.558238
PDy1	−3.674945	−3.674945	−2.23053	−2.23053
PDy2	0.285134	0.285134	0.090785	0.090785
PDy3	−2.494252	−2.494252	−5.71836	−5.71836
PEy1	−0.078785	−0.078785	−0.40009	−0.40009
PEy2	0.245086	0.245086	0.569694	0.569694
PEy3	−0.382274	−0.382274	−0.26276	−0.26276
PEy4	−6.25570332	−6.25570332	−29.3487	−29.3487
PKy1	−41.7228113	−41.7228113	−28.2448	−28.2448
PKy2	2.11293838	2.11293838	1.331304	1.331304
PKy3	0.150080764	0.150080764	0.255683	0.255683
PHy1	0.00711	0.00711	0.00847	0.00847
PHy2	−0.000509	−0.000509	0.000594	0.000594
PHy3	0.049069131	0.049069131	0.042	0.042
PVy1	0.00734	0.00734	0.0262	0.0262
PVy2	−0.0778	−0.0778	−0.0791	−0.0791
PVy3	−0.0641	−0.0641	−0.08552	−0.08552
PVy4	−0.6978041	−0.6978041	−0.44481	−0.44481
λFzo	1	1	1	1
$\lambda \mu y$	1	1	1	1
$\lambda Ky\alpha$	1	1	1	1
λCy	1	1	1	1
λEy	1	1	1	1
λHy	1	1	1	1
λVy	1	1	1	1
$\lambda \gamma y$	1	1	1	1
$\lambda Ky\gamma$	1	1	1	1
Wheel load (N)	381.0	1233.0	628.0	1682.00
Downforce (N)	0	0	0	0
Fz – normal force (N)	381.0	1233.0	628.0	1682.0
Dfz	−0.84408583	−0.49542738	−0.83687824	−0.563103833
Fz0 – nominal load	2443.652224	2443.652224	3849.885	3849.885
α – slip angle (deg)	2.85	2.85	2.25	2.25
α – slip angle (rad)	0.049741884	0.049741884	0.039269908	0.039269908
γ – camber (deg)	0	0	0	0
γ – camber (rad)	0	0	0	0
SHy	0.00753964	0.007362173	0.007972894	0.008135516
SVy	27.81676346	56.57528065	58.02535942	118.9872252
αy	0.057281523	0.057104056	0.047242802	0.047405424
γy	0	0	0	0
Cy	0.324013	0.324013	0.558238	0.558238
μy	−3.91562257	−3.81620819	−2.30650599	−2.281651381
Dy	−1491.8522	−4705.3847	−1448.48576	−3837.737624
Ey	−0.39485848	−0.27674137	−1.10725681	−0.91030711
$Ky\alpha$	−14965.2297	−46067.4414	−26253.0203	−64431.23134
By	30.95958658	30.21596863	32.46726068	30.07473126
Fy0 (N)	−500.91942	−1569.84708	−797.813558	−2048.904262
Normalised Fy	−1.31	−1.27	−1.27	−1.22
Sum		−2071		−2847

148

2.85° as opposed to 3.4° at the front and 2.25° compared to 2.8° at the rear. However the difference of 0.6° (mild understeer) is the same. It is clear that the Pacejka parameters could be better.

Handling curves

It is important to consider the understeer/oversteer balance of a car over the full range of lateral *g* forces. The process described in *Example 5.3* therefore needs to be repeated in increments of lateral *g* up to the peak value. A good method of representing the results is in the form of a handling curve – *Figure 5.16*. The vertical axis shows increments of lateral *g* and the horizontal axis shows the difference in slip angle between the front and rear wheels. The data point for *Example 5.3* is shown (1.25*g*, 0.6°) and the solid line illustrates a typical complete driving curve. It can be seen that in this case it stays on the understeer side of the chart. The curve effectively indicates what the driver must do with the steering wheel to keep the car on the desired radius. In this case it indicates that the steering angle must be gradually increased as lateral *g* forces build, until, when close to the peak, it is increased markedly to avoid an understeer slide. A rapid change near peak grip is common. The lower part of the curve is dominated by the relative slopes of the front and rear tyre test curves – i.e. the **cornering stiffnesses**, whereas the region at maximum *g* is dominated by the location of the peak values. Close to peak grip small changes in load produce large changes in slip angle and so a skilled driver, instead of increasing the steering angle, has the option of reducing rear grip by applying additional drive to the rear wheels.

The dashed line also shows a possible outcome from the analysis for a car where the front and rear roll rates are a shade more biased towards the rear. Here the car exhibits understeer over most of the range but then switches to oversteer at the limit. Unless the driver can react very quickly and apply negative lock, an oversteer spin will result. This leads to a slightly more precise definition of understeer/oversteer. Up to now we have indicated that a car is in an understeer condition if the front slip angles are greater than those at the rear. More formally it is related to the slope of the handling curve. Thus the line shown at the top of *Figure 5.16* indicates the neutral point that divides understeer from oversteer. It corresponds to the point where the driver has to reverse the direction of the steering wheel.

If the resulting handling curve is not to the designer's liking, there are a range of measures available to change the understeer/oversteer balance – *Table 5.3* lists those to reduce oversteer. To reduce understeer apply the opposite. Some changes are appropriate at the design stage and others can be applied in the pits. Some are more desirable than others as they have the effect of strengthening the 'weak' end of the car rather than just weakening the 'strong' end.

Figure 5.16
Handling curves

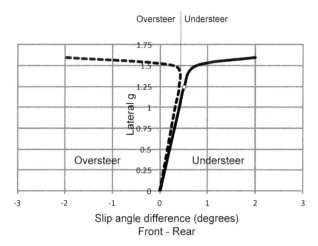

Recent F1 cars were known to suffer an oversteer balance problem largely owing to overloaded rear control tyres. This resulted in the use of an exceptionally stiff front suspension and ballast weights moved as far forward as possible. As a result of the almost rigid front suspension, some drivers lifted a wheel off the track in slow corners where there was insufficient downforce – *Figure 5.17*.

Table 5.3 Measures to reduce oversteer

Measure	Comment
Use wider tyres at the rear	Not allowed where formula specifies control tyres
Increase the roll stiffness at the front axle	Either by increasing spring stiffness, bell-crank ratio or ideally anti-roll bar stiffness
Move centre of mass towards the front	Reduces rear grip but reduces required rear grip force by more. Good if car is ballasted
Apply less front wing downforce	Speed dependent so will work less in slow corners
Raise front ride height or lower rear	Liable to upset underbody aerodynamics
Raise or lower front tyre pressures from optimum	Easy desperate measure!

5.3.3 Jacking

In *section 5.3.1* we saw how to calculate individual wheel loads on a car when subjected to cornering forces. *Step 3* was concerned with calculating those loads that are transferred through the suspension links. It is often assumed that because the load is, in this case, applied at the roll centre it has no effect on the load in the springs. This is not the case for cars where the roll centre is above or below ground level. It is also interesting that the entire calculation in *section 5.3.1* did not require any assumptions to be made about the distribution of lateral grip. *Figure 5.18* represents either end of a car where the cornering force, F_H, has been calculated as in *section 5.3.1, step 3*.

Chapter 5 Tyres and balance

Figure 5.17
Charles Pic lifting an inner wheel at the 2013 German Grand Prix (reproduced with the kind permission of Caterham F1 Team)

Load transfer $\quad \Delta W = \pm F_H h_{rc}/T$

$\Sigma H = 0 \qquad F_H = F_{inner} + F_{outer}$

If we start by making the (unrealistic) assumption that the lateral grip force at the wheels is divided equally between the inner and outer wheels:

$$F_{inner} = F_{outer} = F_H/2$$

ΣM about roll centre for the outer wheel (inner wheel gives the same result):

Clockwise moment $= F_{outer} \times h_{rc} = F_H h_{rc}/2$

Anticlockwise moment $= F_H h_{rc}/T \times T/2 = F_H h_{rc}/2$

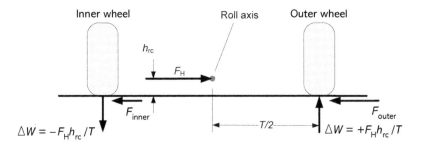

Figure 5.18
Calculating jacking during cornering

Hence the clockwise and anticlockwise moments are equal and the wheel is in moment equilibrium. However if we now choose a more realistic distribution of lateral grip, this equilibrium will be upset. Clearly, grip from the heavily loaded outside wheel, F_{outer}, will be greater than that from the lightly loaded inner, F_{inner}. Once the inner and outer grip forces are assumed to be unequal it can be shown that:

151

Out of balance moment at each wheel = $(F_{outer} - F_{inner})h_{rc}/2$

This is anticlockwise at the inner wheel and clockwise at the outer wheel. Equilibrium must be restored by a change in load in the springs:

$$\text{Change in load at wheel} = (F_{outer} - F_{inner})h_{rc}/(2 \times T/2)$$
$$= (F_{outer} - F_{inner})h_{rc}/T$$

For both the inner and outer wheel the springs will be relieved of load, causing the chassis to rise relative to the wheels. If the wheel centre rate is K_W:

$$\text{Jacking displacement} = \mathbf{(F_{outer} - F_{inner})h_{rc}/TK_W} \quad [5.30]$$

At the limit, as shown in *Figure 5.17*, all of the lateral grip, F_H, may be resisted by the outer wheel.

$$\text{Jacking displacement becomes} = \mathbf{F_H h_{rc}/TK_W} \quad [5.31]$$

EXAMPLE 5.4

For the car in *Examples 5.2* and *5.3*, estimate rear axle jacking in mm when cornering at 1.25 lateral g, given that the wheel centre rate is 41 N/mm.

From *Table 5.2*, approximate rear wheel grip forces from Pacejka (2.25° slip):

$$F_{outer} = 2073 \text{ N}$$
$$F_{inner} = 798 \text{ N}$$

Also height of roll centre, h_{rc} = 77 mm

Track, T_R = 1500 mm

From *equation [5.30]* Jacking = $(F_{outer} - F_{inner})h_{rc}/TK_W$
$$= (2073 - 798) \times 77/(1500 \times 41)$$
$$= \mathbf{1.6 \text{ mm}} \uparrow$$

SUMMARY OF KEY POINTS FROM CHAPTER 5

1. Traditional **bias-ply** tyres provide less ultimate grip than **radial-ply** tyres but the latter fall off more quickly after the peak, making them less forgiving.
2. When cornering, the difference between the direction of travel and the longitudinal axis of a tyre is known as the **slip angle** and it is caused by distortion of the tyre contact patch. The slip angle increases with cornering force. A tyre with a slip angle produces **drag**.
3. If the slip angles of the front tyres are the same as those at the rear, the car is said to have **neutral handling**. If the front slip angles are greater than the rears, the car is **understeering**. If the rear slip angles are greater than the fronts, the car is **oversteering**.
4. Understeer is easier for the driver to control than oversteer which often results in a spin, however drivers differ in the handling balance that they prefer.
5. During acceleration a driven wheel will rotate faster than a free-rolling wheel and this is expressed as a **slip ratio**. A braked wheel will rotate more slowly.
6. Tyre testing produces vital data on the relationship between tyre grip and slip angle or slip ratio. This can be used to estimate the handling balance of a car at the design stage.
7. Testing data can also be represented in the form of a mathematical model such as the Pacejka Magic Tyre Model, and this means that the design process can be computerised.
8. The understeer/oversteer balance over the full range of lateral g forces can be represented in the form of a **handling curve**.

6 Front wheel assembly and steering

LEARNING OUTCOMES

At the end of this chapter:
- You will know which components are involved in the front wheel assembly and how they are packaged
- You will understand that it is important to minimise the unsprung mass
- You will be able to define various aspects of front wheel geometry
- You will understand racing car steering systems and how to avoid problems such as bump steer
- You will be able to specify wheel bearings
- You will learn how to evaluate the loads on wheel uprights to facilitate effective cornering and braking

6.1 Introduction

This chapter is concerned with the design of the front wheel assemblies and includes wheel alignment, bearings and steering. *Figure 6.1* illustrates the various components involved. The **upright** is the main structural component which supports the **axle bearings** (not shown). It also supports the brake calliper and transmits all the forces from the wheel to the suspension members. The steering tie rod connects the steering rack to the steering arm which is connected to the upright. When cornering, the whole assembly pivots about spherical bearings at the ends of the upper and lower wishbones.

It is important to minimise the weight of all wheel assembly components which form **unsprung mass**. This facilitates good control of the oscillating wheel by the spring/damper system without unduly unsettling the sprung chassis. An analogy is to imagine standing in the back of a pickup truck trying to hold a large hammer horizontal as the truck moves over rough ground. Clearly as the mass of the hammer head grows, increasingly large forces are required to hold it steady.

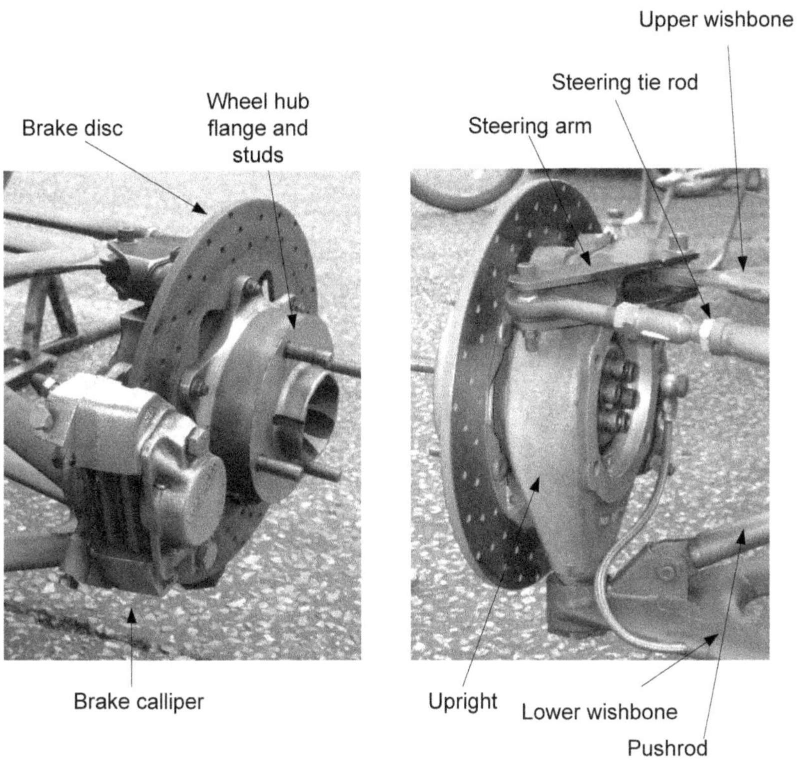

Chapter 6 **Front wheel assembly and steering**

Figure 6.1
Front wheel assembly – Van Diemen RF99 Formula Ford

Labels: Brake disc; Wheel hub flange and studs; Upper wishbone; Steering tie rod; Steering arm; Brake calliper; Upright; Lower wishbone; Pushrod

Packaging of the components inside the wheel presents an interesting challenge. If off-the-shelf wheels are to be used the process starts with a cross-section of the wheel profile as given in *Figure 6.2*, which shows a typical F3 wheel. The first component to be added is the wheel hub flange. (Unlike *Figure 6.1*, *Figure 6.2* shows a centre-lock wheel, where the wheel is retained by a single nut and the brake torque resisted by shear-pegs.) The brake disc and calliper are added next. Generally the calliper is positioned as far into the wheel as possible and as close to the rim as possible so as to maximise the diameter of the disc. The wheel will deform under load so a minimum gap of a few millimetres between the rim and the calliper is required. The upright is now introduced. The lower spherical-bearing ball-joint is positioned as far into the wheel and as low as possible. Maximising the vertical separation between the upper and lower ball-joints minimises the forces in the wishbones. Great care must be taken to ensure there is adequate clearance between the wheel rim and the suspension and steering components under all steering angles and suspension movements. The position of the upper ball-joint can now be fixed. This defines two important parameters of front-wheel geometry – ***steering axis inclination*** and ***scrub radius***.

6.2 Front wheel geometry

6.2.1 Steering axis inclination and scrub radius

The **steering axis inclination**, traditionally known as the **kingpin inclination (KPI)**, is the *front-view* angle between the centre-line of the wheel and the line passing between the upper and lower ball-joints (the steering axis). The **scrub radius** or **kingpin offset** is the lateral distance between these two lines at road level. An excessive scrub radius means that any out-of-balance forces acting on the front wheels cause a moment to be transmitted to the steering system. Thus if one front wheel hits an obstruction, or road conditions cause uneven braking, the driver would have to resist a sudden tug at the steering wheel. The aim, therefore, is to keep the scrub radius to a value below say 40 mm. It can be seen that, in the example shown in *Figure 6.2*, this requires significant kingpin inclination.

There are, however, good reasons why the value of kingpin inclination should be kept as low as possible. As the wheel rotates around the steering axis during cornering, KPI produces the following effects:

1. The camber values of the front wheels change. The heavily loaded outer wheel gains adverse positive camber as the steering angle increases. (The lightly loaded inner wheel also gains positive camber, but this is desir-

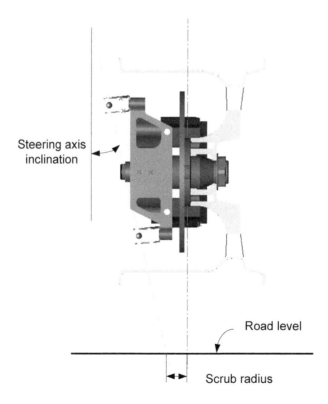

Figure 6.2
Front wheel assembly

able.) The actual change in camber, Δγ, as a result of camber angle $θ_k$ at steering angle δ is given by:

$$Δγ = θ_k + \cos^{-1}(\sin θ_k \cos δ) - 90° \qquad [6.1]$$

2. The tyre contact patch moves vertically – known as **lift** or **jacking** and, as we saw in *Chapter 5*, this will change lateral load transfer during cornering.

6.2.2 Caster angle and caster trail

Two further important front-wheel geometry parameters are ***caster angle*** and ***caster trail*** which can also be referred to as **mechanical trail**. Caster angle is the *side-view* angle between the centre-line of the wheel and the line passing between the upper and lower ball-joints (the steering axis). Caster trail is the longitudinal distance between these two lines at road level – as shown in *Figure 6.3*. Caster trail is the primary mechanism for providing the self-centring effect to the steering wheel. An element of caster trail is also required for straight-line stability. Racing car steering systems can contain free-play as a result of gear backlash and wear. A car with inadequate caster trail is liable to oscillate, in a weaving fashion, on straights as the driver repeatedly corrects for backlash. A minimum caster trail value of about 15 mm appears adequate for this purpose. Significantly larger caster trail values should be avoided as the resulting self-centring effect can dominate the effect of **pneumatic trail** (see *section 5.2.2*) which provides the driver with valuable steering 'feel'.

The use of caster angle can counter some of the adverse effects of KPI on wheel camber. As the wheel rotates around the steering axis during cornering, caster angle produces the following effects:

1. The heavily loaded outer wheel gains negative camber as the steering angle increases – i.e. the opposite effect to KPI. (The lightly loaded inner

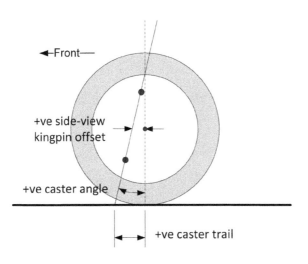

Figure 6.3
Caster angle and caster trail

wheel gains additional positive camber.) The actual change in camber, $\Delta\gamma$, as a result of caster angle θ_c at steering angle δ is given by:

$$\Delta\gamma = \cos^{-1}(\sin\theta_c \sin\delta) - 90° \qquad [6.2]$$

Equation [6.2] results in a negative value of $\Delta\gamma$ for positive values of δ. It is clear that, if desired, an optimum amount of caster angle can be added so that the combined effects of *equations [6.1] and [6.2]* leave the outer wheel with virtually no change in camber during cornering.

2. Caster angle, combined with KPI, increases lift significantly. The lightly loaded inner wheel contact patch moves downwards and the outer wheel contact patch less so. This means that the four-wheel loads will change. The loads on the diagonal formed by the front inner wheel and the rear outer wheel will increase and those on the other diagonal will decrease. This causes less load transfer at the front and more at the rear and, because of the phenomenon of tyre sensitivity, this will strengthen grip at the front and weaken it at the rear. Hence the balance of the car will move towards oversteer as the steering angle increases. The significance of lift depends to some extent on the wheel stiffness of the car. Hence 2 mm of relative lift on a car with a wheel stiffness of 30 N/mm will change wheel loads by about ±15 N. Lift is also responsible for some of the self-centring effect when steering, i.e. the tendency of the steering wheel to return to the straight-ahead position when released. (Racing karts, without a rear differential, use large caster angles to lift the inner rear wheel, so that it spins when cornering!)

It is clear from *Figure 6.3* that caster trail is made up of two components – that which results from caster angle and that which results from side-view kingpin offset. Side-view kingpin offset is zero if the kingpin axis passes through the wheel centre. It has been known for designers to adopt a large caster angle to get the benefits of negative camber change, but to combine

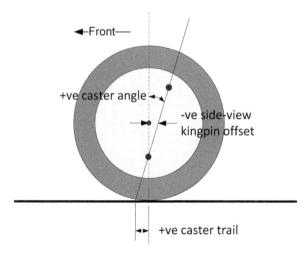

Figure 6.4
Combining a large caster angle with negative kingpin offset

Chapter 6 Front wheel assembly and steering

Table 6.1 Camber change on outer wheel with caster angle and KPI

Change of camber with caster angle and kingpin inclination	
Caster (deg)	4
King PI (deg)	11
Steer. inc.	5

Camber change with caster angle and KPI

(graph showing camber change vs steering angle, with three curves: Camber change (caster), Camber change (KPI), Camber change (total))

Steering angle	Camber change (caster)	Camber change (KPI)	Camber change (total)
0	0.00	0.00	0.00
5	−0.35	0.04	−0.31
10	−0.69	0.17	−0.53
15	−1.04	0.37	−0.66
20	−1.37	0.66	−0.70
25	−1.69	1.03	−0.66
30	−2.00	1.47	−0.53
35	−2.29	1.99	−0.30

this with negative side-view kingpin offset to keep the total caster trail reasonable. This is shown in *Figure 6.4*.

A simple spreadsheet has been prepared that plots the values of camber change from *equations [6.1] and [6.2]* to show the combined effect of KPI and caster angle. The results are shown in *Table 6.1* and the spreadsheet can be downloaded from www.palgrave.com/companion/Seward-Race-Car-Design.

6.2.3 Static toe

The final element of front wheel geometry is **static toe**. This takes the form of **toe-in** or **toe-out** and is shown exaggerated in *Figure 6.5*. In practice, very small amounts of toe can significantly alter the feel of a car. Toe-in, where the

Figure 6.5
Front wheel toe-in and toe-out

Toe-in   Toe-out

centre-line of the wheels converge at the front of the car, produces a more stable car on straights, whereas toe-out produces a more lively car in corners.

The most plausible explanation for this toe effect concerns tyre drag. *Figure 6.6* shows the same two cars just starting to turn left. In the case of the toe-in car, the driver has turned the steering wheel just enough to make the left-hand wheel point straight-ahead. The effect of lateral load transfer has not yet become established. The right-hand wheel is pointing to the left and starts to develop a lateral force perpendicular to the plane of the wheel. This force can be split into a horizontal component which tries to turn the car and a longitudinal drag force which tries to pull the car out of the turn. It could be likened to a sticking brake pulling the car to the right. With the toe-out car, the left-hand wheel is turned and the drag force adds to the turning force to pull the car into the turn. The car could be said to have 'good turn-in'.

Toe can be measured either in degrees or, more commonly, a toe-gauge is used to measure the distance between the inside of the wheel rims both at the front and rear of the wheel. The difference between these two measurements is the **total toe** (twice the normally specified toe) and rarely exceeds say 3 mm which corresponds to about 0.25°.

6.3 Steering

Racing car steering systems invariably use a **rack and pinion** to convert the rotary motion of the steering wheel into linear motion of the steering tie rod.

Chapter 6 **Front wheel assembly and steering**

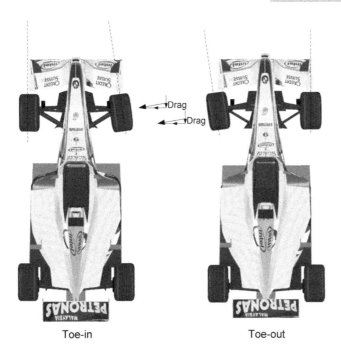

Figure 6.6
Initial cornering with toe-in and toe-out

The usual requirement is for a *fast* rack (i.e. highly geared) so that the driver can reach full lock with less than half a turn each way on the wheel; however this can result in high forces at the steering wheel. The **steering arm** is rigidly connected to the wheel upright and is the lever that turns the wheel about the steering axis. Both the length of the steering arm and the gearing of the rack and pinion determine the relationship between turns of the steering wheel and rotation of the wheel.

F1 cars now use **power steering** to reduce driver fatigue and allow the use of very small steering wheels. Most small single-seaters cope with purely manual steering partly because the duration of most races is relatively short.

6.3.1 Parallel or Ackermann steering?

Figure 6.7 shows three steering arrangements that deliver approximately **parallel steering**, i.e. both front wheels will turn through roughly the same angle. The common factor between them is that the steering tie rod is perpendicular to the steering arm. We know that when a car corners the outer wheel describes a larger radius than the inner, therefore with parallel steering, particularly when moving slowly, the two tyres will fight each other and scrub.

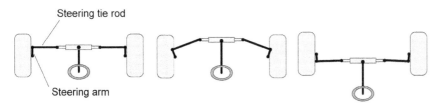

Figure 6.7
Parallel steering systems

161

This is not really a problem at speed when slip angles have developed and load transfer has taken place. If a car has parallel steering the slip angle of the inner wheel will be less than that of the outer.

It is an easy matter to arrange the steering mechanism to turn the inner wheel through a greater angle to accommodate the smaller turn radius, and this is known as **Ackermann steering**. *Figure 6.8* shows a simple method of achieving a good approximation to Ackermann. The steering arms intersect at the rear axle line. This assumes that the steering tie rod is perpendicular to the car centre-line. If this is not the case, the indicated angle between the steering tie rod and steering arm must be preserved. If the steering arms intersect behind the rear axle line, part-Ackermann exists. If a car has full Ackermann the slip angles of both front tyres will be the same during cornering. The development of different front steering angles as the steering wheel is turned is **dynamic toe**, which is added to the small amount of static toe described above.

The question is – when cornering at the limit, what is the best direction to point the inner wheel? It is a fact that for most tyres, at low vertical loads, peak grip occurs at a lower slip angle than for high vertical loads – typically the difference might be about 3°. It can therefore be argued that the inner wheel should be turned about 3° less than that indicated by full Ackermann. A simple spreadsheet is available to calculate the Ackermann difference between the front wheel steering angles as well as the outer turn radius for the car. This can be downloaded from www.palgrave.com/companion/Seward-Race-Car-Design. The results are as shown in *Table 6.2*.

Figure 6.8
Ackermann steering

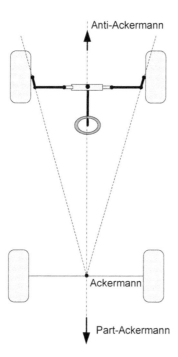

Table 6.2 Calculation of Ackermann angles and turning radius

Ackermann calculator						
F. track (mm)	1200					
W/base (mm)	1600					
Outer wheel angle°	0	5	10	15	20	25
Inner wheel angle°	0.0	5.3	11.5	18.5	26.6	35.6
Outer turn rad. (m)		18.4	9.2	6.2	4.7	3.8

It can be concluded from the spreadsheet that:

- There is very little difference between the inner and outer wheels at up to 5° of steering angle.
- Turn radius is highly dependent upon wheelbase. The car dimensions shown in *Table 6.2* are typical of Formula SAE/Student cars which are required to negotiate corners with an outside radius as small as 4.5 m. Hence such cars need just over 20° of lock for the outer wheel. With a typical racing car wheelbase of 2.5 m this goes up to 30°, which can be tricky to achieve – hence Formula SAE/Student cars need to be short.

Typical full-sized circuit racing cars are rarely expected to turn radii less than 10 m so a maximum steering angle of around 10° is required on the track. More may be needed for navigation in the pits however. At 10° on the outer wheel it can be seen that there is 11.5° on the inner wheel at full Ackermann. From the above, for peak grip, we may want 3° less on the inner wheel, i.e. 8.5°. Parallel steering would deliver 10°. Hence some designers argue for **reverse** or **anti-Ackermann**, i.e. the inner wheel turning say 1.5° *less* than the outer. To achieve this the intersection point for the steering arms is a wheelbase length *in front of* the car.

Anti-Ackermann geometry makes navigation at slow speeds in the pits difficult, so many designers settle for parallel steering. Also for tyres with slip angle curves that exhibit a long flat peak, the benefits of anti-Ackermann are negligible.

6.3.2 Bump steer

Bump steer or **ride steer** is said to occur when the front wheels rotate about the steering axis as the suspension rises or falls relative to the chassis. It is generally a bad thing and should be minimised (although some designers claim benefits from building-in small amounts of dynamic toe from bump steer as a driver brakes before a corner). It is caused by the distance between the steering tie rod joints changing as the suspension rises and falls. Bump steer is minimised if the steering tie rod lies on a line which points to the

Race car design

Figure 6.9
Avoiding bump steer

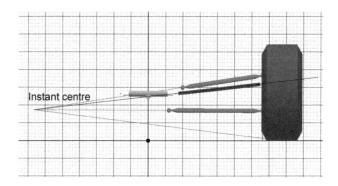

suspension link instant centre as shown in *Figure 6.9*. The situation gets a little more complicated if the suspension links contain anti-dive geometry. Bump steer is sensitive to small adjustments in the height of the steering rack. Suspension design software such as SusProg calculates bump steer and so an effective design approach is to adjust the rack height until the software indicates minimal bump steer effect.

6.4 Axle design and bearings

The design of virtually all wheel assembly components starts with an assessment of loads at the tyre/road interface for different load cases as shown in *Figure 3.14*.

6.4.1 Axles

The governing load case for axle design is maximum cornering which should include the effect of extra grip from aerodynamic downforce. The forces W_{vert} and W_{lat} are shown in *Figure 6.10*.

The maximum bending moment in the axle occurs at the centre of the outer bearing:

$$M_{axle} = W_{lat}R_r - W_{vert}l_2$$

From *equation [2.3]*:

$$\text{Elastic modulus, } Z = \frac{1.5 \times M_{axle}}{\sigma_y}$$

where 1.5 = material factor-of-safety and for a solid circular shaft:

$$Z = \frac{\pi r^3}{4}$$

Chapter 6 Front wheel assembly and steering

Figure 6.10
Front axle and wheel bearings

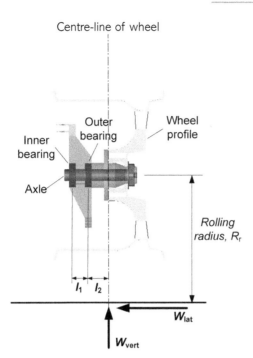

It pays to use a good high-strength alloy steel as this reduces the size and hence the mass of the axle, bearings and upright. BS EN 24T or 817 M40 has a yield stress of 650 N/mm² and is machineable.

EXAMPLE 6.1

Determine the required diameter of an EN 24T alloy steel axle given the following data:

W_{lat} = 4275 N, W_{vert} = 2850 N, bearing spacing, l_1 = 44 mm, distance l_2 = 53 mm, rolling radius, R_r = 270 mm.

$$M_{axle} = W_{lat}R_r - W_{vert}l_2 = 4275 \times 270 - 2850 \times 53$$
$$= 1\,003\,200 \text{ Nmm}$$

$$\text{Elastic modulus, } Z = \frac{1.5 \times M_{axle}}{\sigma_y} = \frac{1.5 \times 1\,003\,200}{650}$$
$$= 2315 \text{ mm}^3$$

Try a 30 mm diameter axle:

$$Z = \frac{\pi r^3}{4} = \frac{\pi \times 15^3}{4}$$
$$= 2651 \text{ mm}^3 > 2315$$

Conclusion:

Use a 30 mm diameter EN 24T axle

6.4.2 Front wheel bearings

Front wheel bearings need to resist combined radial and axial loads and generally take the form of a pair of either tapered roller bearings or angular contact ball bearings. A **tapered roller bearing** tends to have a higher load capacity and be slightly cheaper and heavier than an equivalent **angular contact ball bearing**. The latter also generates less friction. The design procedure is very similar for both types. They must be properly housed or located and are normally subject to a small pre-load via a locking nut. *Figure 6.11* shows a pair of angular contact ball bearings and how the angle of thrust increases the effective spacing of the bearings. This helps to resist bending in the shaft but creates additional axial forces. There are two main design cases – static loads and dynamic loads. The design procedure for combined radial and axial loads can be quite complex, however, in our case, the axial loads are normally small enough to ignore as in the following simplified approach.

Figure 6.11 Front wheel bearing design

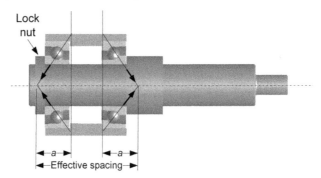

Design for static loads

This is not simply concerned with static loads but is a check to ensure that the absolute maximum load expected on the bearing will not cause any permanent deformation or yielding of the bearing race or rolling elements.

$$\text{Static safety factor, } s_0 = \frac{C_0}{P_0} \qquad [6.3]$$

where C_0 = basic static load rating from bearing data sheet
 P_0 = maximum radial load on bearing (in our case).

For angular contact ball bearings $s_0 \geq 1.0$ for quiet running but $s_0 \geq 0.5$ if some rumble can be tolerated. The equivalent figures for tapered roller bearings are 1.5 and 1.0.

Design for dynamic loads

This is a fatigue check to ensure that the working life of the bearing is adequate. It is not based on peak bearing loads but on a **mean equivalent dynamic load** derived from consideration of a typical loading spectrum throughout the life of the bearing. The procedure starts by evaluating the bearing radial load during different operations such as cornering, braking, accelerating ... etc. – say $P_1, P_2, P_3 \ldots$. The proportion of time spent on each operation is then estimated – say $T_1, T_2, T_3 \ldots$.

Mean equivalent dynamic load, $P_m = \sqrt[3]{(P_1^3 T_1 + P_2^3 T_2 + P_3^3 T_3 \ldots)}$ [6.4]

P_m is then compared to the basic dynamic load rating, C_r, from the bearing data sheet. The basic dynamic load assumes one million revolutions before fatigue failure.

$$\text{Estimated fatigue life} = \left(\frac{C_r}{P_m}\right)^3 \times 10^6 \text{ cycles} \quad [6.5]$$

EXAMPLE 6.2

Specify suitable angular contact ball bearings for a car with the following maximum front wheel loads:

Load case	Vertical load (kN)	Lateral load (kN)	Longitudinal load (kN)
Cornering	2.850	4.275	–
Braking	2.033	–	3.050

Axle dia. = 30 mm, bearing spacing l_1 = 44 mm, distance l_2 = 53 mm (see *Figure 6.10*), rolling radius R_r = 270 mm.

Try angular contact ball bearing 7206 BEP – from manufacturer's (SKF) data sheet:

$$\begin{aligned}
\text{Dimension } a &= 27.3 \text{ mm (see } Figures\ 6.11 \text{ and } 6.12) \\
\text{Width, } B &= 16 \text{ mm} \\
\text{Basic static load, } C_0 &= 14.3 \text{ kN} \\
\text{Basic dynamic load, } C_r &= 22.5 \text{ kN} \\
\text{Effective bearing spacing} &= 44 - 16 + (2 \times 27.3) = 82.6 \text{ mm}
\end{aligned}$$

(*Note* – the vertical wheel loads could be reduced by the weight of the wheel, tyre, axle and hub as these are transmitted directly to the ground without passing through the bearings – however this will be ignored here.

Case 1 – cornering

For radial load on inner bearing take moments about *Y*:
$$-(2.850 \times 35.7) + (4.275 \times 270) = P_{1i} \times 82.6$$
$$P_{1i} = \mathbf{12.7 \text{ kN}}$$

Race car design

Figure 6.12
Front wheel bearing dimensions

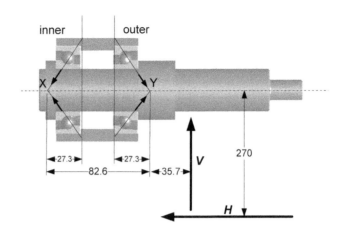

For radial load on outer bearing take moments about X:

$$-(2.850 \times 118.3) + (4.275 \times 270) = P_{1o} \times 82.6$$

$$P_{1o} = \mathbf{9.9 \text{ kN}}$$

Case 2 – braking

Inner bearing vertical load	= 2.033 × 35.7/82.6	= 0.88 kN
Inner bearing longitudinal load	= 3.050 × 35.7/82.6	= 1.32 kN
Inner bearing resultant radial load, P_{2i}	= √(0.88² + 1.32²)	= **1.6 kN**
Outer bearing vertical load	= 2.033 × 118.3/82.6	= 2.91 kN
Outer bearing longitudinal load	= 3.050 × 118.3/82.6	= 4.38 kN
Outer bearing resultant radial load, P_{2o}	= √(2.91² + 4.38²)	= **5.3 kN**

Check static loads

From above, maximum radial load P_{1i} = 12.7 kN

Equation [6.3]   Static safety factor, $s_0 = \dfrac{C_0}{P_0} = \dfrac{14.3}{12.7} = 1.1 > 1.0$ ✓

Check dynamic loads

Dynamic load profile

Activity	Assumed % time	Inner bearing load	Outer bearing load
Turning right	30	12.7	9.9
Turning left	20	3*	4*
Braking	15	1.6	5.3
Accelerating	25	0.8*	1.6*
Cruising	10	1*	2*

* estimated

Equation [6.4]   Mean equivalent dynamic load, P_m

$$= \sqrt[3]{(P_1^3 T_1 + P_2^3 T_2 + P_3^3 T_3 \ldots)}$$

Inner bearing, P_{mi}

$$= \sqrt[3]{[(12.7^3 \times 0.3) + (3.0^3 \times 0.2) + (1.6^3 \times 0.15) + (0.8^3 \times 0.25) + (1.0^3 \times 0.1)]}$$

$$= 8.5 \text{ kN}$$

Outer bearing, P_{mo}

$$= \sqrt[3]{[(9.9^3 \times 0.3) + (4.0^3 \times 0.2) + (5.3^3 \times 0.15) + (1.6^3 \times 0.25) + (2.0^3 \times 0.1)]}$$

$$= 6.9 \text{ kN}$$

Equation [6.5]

$$\text{Estimated fatigue life} = \left(\frac{G_r}{P_m}\right)^3 \times 10^6 \text{ cycles}$$

$$\text{For inner bearing life} = \left(\frac{22.5}{8.5}\right) \times 10^6 \text{ cycles} = 18.5 \times 10^6 \text{ cycles}$$

Assuming say 5000 km of racing:

Circumference of wheel = π × 0.540 = 1.7 m

Number of revolutions = $5000 \times 10^3 / 1.7$ = 3.0×10^6 < 18.5×10^6 ✓

Use angular contact ball bearing 7206 BEP

6.5 Upright design and analysis

The main structural element of the wheel assembly is the **upright**. This transfers load from the axle bearings and brake calliper to the suspension wishbones. A wide range of manufacturing techniques are available including aluminium casting (*Figure 6.1*) and fabricated steel; however the current favoured approach is numerical control (NC) machining from good-quality aluminium alloy such as 7075-T6. Although not the cheapest method, this probably produces the lightest solution to this important element of unsprung mass. As with the chassis frame, the best structural approach is triangulated members between the load points as shown in *Figure 6.13*. In this case the top of the upright connects directly to the wishbone, whereas at the bottom a bolted channel connects to the lower upright and steering tie rod. A bracket is provided for a lug-mounted brake calliper. It can be seen that, in this case, triangulated members connect the principal load points to the central circular bearing housing.

Figure 6.13
Typical triangulated front upright

Race car design

6.5.1 Load cases

The front upright should be checked for two main load cases – maximum cornering and maximum braking.

Maximum cornering

As usual, the analysis starts with an assessment of loads at the tyre/road interface as shown in *Figure 3.14*. For analysis purposes it is assumed that the upright is restrained at the upper and lower connection points. It is important that the structure is not over-restrained. The ends of the wishbones contain spherical bearings that permit rotation and hence 'fixed' supports are not appropriate. The loads are transferred from the wheel to the axle and then applied to the upright via the bearing housings.

Figure 6.14a shows the upright with the maximum cornering loads applied at the contact patch. The bearing loads shown in *Figure 6.14b* are exactly equivalent.

Taking moments about the centre of the inner bearing:

$$F_{outer} = \frac{(W_{lat} \times R_r) - (W_{vert} \times (l_1 + l_2))}{l_1}$$

Figures 6.14a and b
Analysis of front upright for maximum cornering

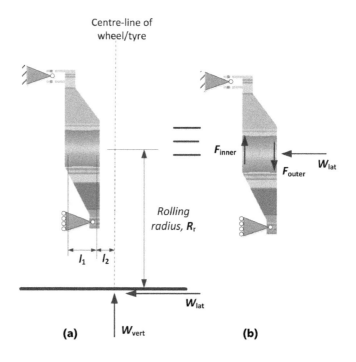

Chapter 6 Front wheel assembly and steering

Summing vertical forces:

$$F_{inner} = F_{outer} + W_{vert}$$

Forces F_{outer} and F_{inner} are applied as bearing forces in the bearing housings. This ideally means a sinusoidal or parabolic distribution over one side of the housing. The force W_{lat} is applied as a uniformly distributed load over the annular step face retaining the bearing.

Maximum braking

Figure 6.15a shows the upright with the maximum braking loads applied at the contact patch. The bearing loads shown in *Figure 6.15b* are exactly equivalent.

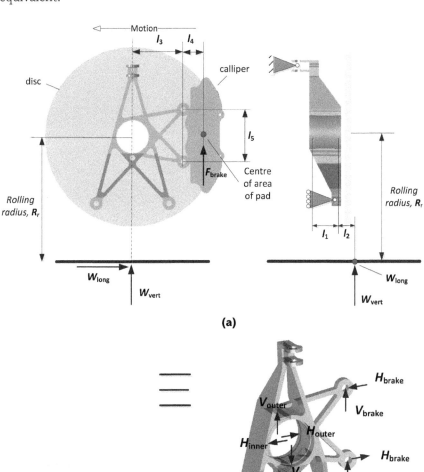

Figures 6.15a and b
Loads on front upright in maximum braking

Taking moments about the centre of the inner bearing:

$$V_{outer} = \frac{W_{vert} \times (l_1 + l_2)}{l_1}$$

$$H_{outer} = \frac{W_{long} \times (l_1 + l_2)}{l_1}$$

Summing vertical forces:

$$V_{inner} = V_{outer} - W_{vert}$$

Summing horizontal forces:

$$H_{inner} = H_{outer} - W_{long}$$

Assuming the wheel is on the point of locking under braking, all of the torque is resisted by the brake calliper. The force will act at the centre of area of the brake pad:

$$F_{brake} = \frac{W_{long} \times R_r}{(l_3 + l_4)}$$

This force can be assumed to be divided equally between the two fixing lugs:

$$V_{brake} = \frac{F_{brake}}{2}$$

In addition because the calliper support lugs are not coincident with the centre of area of the pads there is a moment that causes equal and opposite horizontal forces:

$$H_{brake} = \frac{F_{brake} \times l_4}{l_5}$$

EXAMPLE 6.3

Determine the front upright loads under both maximum cornering and maximum braking for the case considered in *Example 6.2*. The relevant data is repeated below:

Load case	Vertical load (kN)	Lateral load (kN)	Longitudinal load (kN)
Cornering	2.850	4.275	–
Braking	2.033	–	3.050

Bearing spacing l_1 = 44 mm, distance l_2 = 53 mm (see *Figure 6.10*), rolling radius R_r = 270 mm.

In addition:

$l_3 = 70$ mm, $l_4 = 30$ mm, $l_5 = 80$ mm.

Maximum cornering

Taking moments about the centre of the inner bearing:

$$F_{outer} = \frac{(W_{lat} \times R_r) - (W_{vert} \times (l_1 + l_2))}{l_1}$$

$$= \frac{(4.275 \times 270) - (2.850 \times (44 + 53))}{44}$$

$$= \mathbf{19.95 \text{ kN} \uparrow}$$

Summing vertical forces:

$$F_{inner} = F_{outer} + W_{vert}$$

$$= 19.95 + 2.850$$

$$= \mathbf{22.8 \text{ kN} \downarrow}$$

Maximum braking

Taking moments about the centre of the inner bearing:

$$V_{outer} = \frac{W_{vert} \times (l_1 + l_2)}{l_1} = \frac{2.033 \times (44 + 53)}{44} = \mathbf{4.5 \text{ kN}}$$

$$H_{outer} = \frac{W_{long} \times (l_1 + l_2)}{l_1} = \frac{3.050 \times (44 + 53)}{44} = \mathbf{6.7 \text{ kN}}$$

Summing vertical forces:

$$V_{inner} = V_{outer} - W_{vert} = 4.5 - 2.033 = \mathbf{2.5 \text{ kN}}$$

Summing horizontal forces:

$$H_{inner} = H_{outer} - W_{long} = 6.7 - 3.050 = \mathbf{3.7 \text{ kN}}$$

Brake torque:

$$F_{brake} = \frac{W_{long} \times R_r}{l_3 + l_4} = \frac{3.050 \times 270}{(70 + 30)} = \mathbf{11.9 \text{ kN}}$$

$$V_{brake} = \frac{F_{brake}}{2} = \frac{11.9}{2} = \mathbf{6.0 \text{ kN}}$$

$$H_{brake} = \frac{F_{brake} \times l_4}{l_5} = \frac{6.0 \times 30}{80} = \mathbf{2.25 \text{ kN}}$$

6.5.2 Analysis

Whereas it is still possible to carry out simple meaningful hand calculations to determine support reactions and approximate tensile and compressive forces in triangulated upright members, the complex 3-D shape of uprights is generally best analysed by means of a professional finite element package. The usual failure criterion for such an analysis is the von Mises yield criterion which takes into account the 3-D nature of the problem. The von Mises stress is compared to the yield stress of the material and a suitable factor-of-safety of say 1.5, as indicated in *Chapter 2*, is aimed for.

Plates 6(a) and (b) show the results of a finite element analysis for the two load cases covered in *Example 6.3*. The material used for the upright is aluminium alloy 7075 T6 which has a yield strength of 505 N/mm^2. The contour plots show the von Mises stresses. With a factor of safety of 1.5 the von Mises stresses should be less than 505/1.5 = 337 N/mm^2.

Plate 6(a) shows results for the maximum cornering case. The plot indicates that the von Mises stresses are less than 250 N/mm^2 over virtually all of the structure. The stresses are however higher in a small region at the top of the bearing housing ring where they reach a peak of 464.8 N/mm^2. Another contour plot also indicates that the deformation at this point is excessive at 4 mm. Deformation of the upright can adversely affect wheel camber and a maximum deflection of a quarter of this value would be desirable. The designer would probably increase the thickness of this ring at the top by say 2 mm and repeat the analysis.

Plate 6(b) shows results for the maximum braking case. The main body of the upright appears satisfactory but the calliper support bracket shows excessive stresses of 526.6 N/mm^2 in a few small regions. This should be made more robust.

The designer may choose to optimise the structure further by removing material from areas that are consistently under-stressed provided deflection is acceptable.

6.6 Wheel fixings

Wheels are secured to the hubs either by chamfered nuts on multiple studs (*Figure 6.1*), multiple bolts (like modern road cars) or a single **centre-lock** nut (*Figure 6.2*). Centre-lock wheels can be lighter and quicker to remove in pit stops but are notoriously troublesome. Even F1 teams lose centre-lock wheels from time to time. The system consists of a series of drive pegs on the hub that mate with holes in the wheel to transmit torque. A single central nut compresses and retains the wheel on the hub. A lightweight wire clip is used as a failsafe device to stop the nut actually falling off. Traditionally, right-hand-threaded nuts are used on the left-hand side of the car and vice versa.

Several dubious explanations exist to justify this. One of the more plausible is that, because of its polar inertia, a torque is required to accelerate and decelerate the spinning nut. If this torque is greater than the frictional force required to turn the nut, it will move on the threads. A loosely fitting nut may move a thread or two every time the car accelerates and brakes. With the above arrangement of left-hand and right-hand threads the nuts would move in the tightening direction during acceleration and in the loosening direction during braking. However the initial tightening torque of the nut should be much greater than the inertia torque so the nut should never loosen. Other phenomena are at work that involve greater torques. There appears to be a particular problem once the drive peg holes in the wheels start to wear, thus permitting a small amount of relative rotation between the wheel and the hub under braking. This can cause progressive loosening, particularly as the wheel presses against the nut as the driver brakes into a corner. Most users of centre-lock wheels learn that the nuts need to be *really* tight and torques of over 500 Nm are common.

> **SUMMARY OF KEY POINTS FROM CHAPTER 6**
> 1. The position and inclination of the steering axis when viewed from the *front* of the car defines **kingpin inclination** and the **scrub radius**. The scrub radius should not be too big so as to avoid shocks through the steering wheel when a wheel hits a bump.
> 2. The position and inclination of the steering axis when viewed from the side of the car defines **caster angle** and **caster trail**. Some caster trail is necessary for stability but too much can hide steering feel.
> 3. Kingpin inclination and caster angle both affect wheel camber during a turn.
> 4. A small amount of **static toe** can significantly affect the feel of a car.
> 5. Racing cars generally adopt **rack-and-pinion** steering. The initial angle between the steering tie rod and the steering arm determines the degree to which the wheels at each side of the car turn by different amounts – the so-called Ackermann effect. The fact that a lightly loaded tyre achieves peak grip at a lower slip angle than a heavily loaded one means that some designers aim for anti-Ackermann geometry.
> 6. Under **bump steer**, vertical movement of the suspension causes a change in the steering angle. It should generally be avoided by aiming the steering rod at the suspension link instant centre.
> 7. Axles are relatively highly stressed in bending and should be made from good-quality material.
> 8. Front wheel bearings are often either **angular contact ball bearings** or **tapered roller bearings**. These are used in pairs and should be selected for both adequate fatigue life and ultimate load.
> 9. Front **uprights** should be optimised for maximum cornering and maximum braking. The von Mises stresses from a finite element package are compared with the yield stress of the material to ensure an adequate factor of safety.

7 Rear wheel assembly and power transmission

LEARNING OUTCOMES

At the end of this chapter:
- You will know which components are involved in the rear wheel assembly
- You will know about the elements of a car transmission including the clutch, gears and driveshafts
- You will understand the relationship between power and torque and how they influence the optimum gear ratios for racing
- You will know about the types of differential and how they influence the handling behaviour of a car. Also you will be able to calculate the forces required to restrain a differential
- You will be able to specify driveshafts, rear wheel bearings and uprights
- You will learn about driver aids such as launch control

7.1 Introduction

This chapter covers the design of the rear wheel assembly (*Figure 7.1*) and includes the transmission of power to the rear wheels. The fixed **toe control rod** or 'fifth link' replaces the front steering tie rod. Fine adjustment in length is required to set **rear toe**.

The components involved in transmitting the power from the engine are the clutch, gearbox, ***differential*** and ***driveshafts***. In addition, for motorcycle-engined cars, it is usual to retain chain driven front and rear sprocket gears. Each of these will be considered in more detail.

Like with the front assembly, it is important to minimise unsprung mass. Adams (*ref. 1*) explains how reducing the mass of transmission components reduces rotating inertia, which makes more engine power available for accelerating the car.

Chapter 7 **Rear wheel assembly and power transmission**

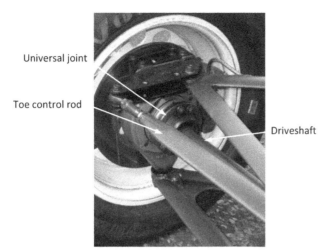

Figure 7.1
Rear wheel assembly – Van Diemen RF99 Formula Ford

7.2 Clutch and gears

Having appropriate gear ratios is vital for good performance. The ratios adopted by manufacturers for road vehicles are generally not ideal for racing. First gear is designed for emergency hill starts and tends to be too low. Top gear is optimised for maximum economy during motorway cruising and is too high. The intermediate ratios may be aimed at minimising the well-publicised 0–60 time! The starting point for specifying gear ratios is the engine power and torque curves. *Figure 7.2* shows a typical example for a restricted 600 cc motorcycle engine of the type commonly used in FSAE/Formula Student. It can be seen that **torque** reaches a peak of 55 Nm at about 9000 rpm whereas

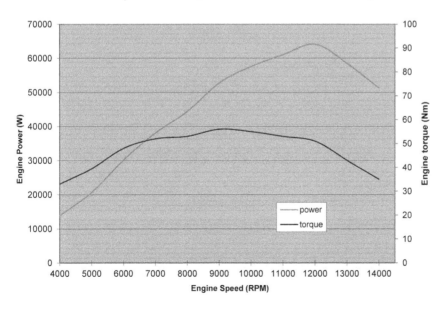

Figure 7.2
Typical power and torque curve – 600 cc restricted motorcycle engine

177

Figure 7.3
Torque transmission in a friction clutch

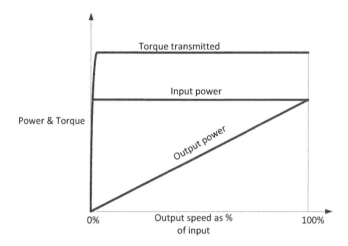

power continues to rise up to a peak of about 64 kW at 12 000 rpm. Optimum ratios for racing depend upon the type of racing and speed of the track, and this often results in a close ratio gearbox where the driver can keep the engine revs close to the peak power point for most of the time. With reference to *Figure 7.2*, this means keeping the engine spinning between say 10 500 and 13 500 rpm.

The relationship between the torque curve and the power curve is simple:

$$\text{Torque} = \text{force} \times \text{radius (Nm)}$$

$$\text{Work done in one revolution} = \text{force} \times 2\pi \times \text{radius} = 2\pi \times \text{torque (Joule)}$$

$$\text{Power at 1 rpm} = \frac{2\pi}{60} \times \text{torque (W)}$$

Hence

$$\textbf{Power} = \frac{2\pi}{60} \times \textbf{torque} \times \textbf{rpm (W)}$$

7.2.1 The clutch

The purpose of the clutch is to disengage the engine from the drivetrain at tickover and to enable the driver to pull-away by changing the relative speed of the engine and the wheels. By partially engaging, or slipping, the clutch the driver can keep the engine speed in the region where it generates sufficient torque to accelerate the car. Once the clutch is fully released (engaged), the gear ratio determines the relationship between the speed of the engine and the speed of the wheels (assuming no wheelspin).

Unlike gears, a partially engaged clutch cannot multiply engine torque; however, rather surprisingly, full engine torque can be transferred through a slipping clutch. *Figure 7.3* shows that for a friction clutch the torque transmitted is independent of the speed differential of the clutch plates.

7.2.2 First gear – getting off-the-line

The essence of getting off-the-line is good clutch and throttle control. The objective is to keep the engine rpm at a value where the engine can supply sufficient torque to generate the maximum traction force at the wheels, but to avoid too much wheelspin. In *Example 1.2* we showed how, by assuming a coefficient of friction, it was possible to calculate longitudinal weight transfer during the early stages of acceleration and hence estimate the peak torque for rear wheels of a particular rolling radius, T_{wheels}. To convert this to the required engine torque, T_{engine}, the wheel torque is divided by the total gear ratio between the engine crankshaft and the wheel driveshafts.

$$\text{Required } T_{engine} = T_{wheels} / \text{gear ratio}$$

EXAMPLE 7.1

In *Example 1.3* we calculated the maximum torque required to spin the driven wheels. A similar calculation was performed for a FSAE/Formula Student car which indicated a peak torque requirement at the wheels of **820 Nm**.

The gear ratios for the proposed engine are as follows:

Primary reduction	1.822
1st	2.833
2nd	2.062
3rd	1.647
4th	1.421
5th	1.272
6th	1.173
Small sprocket teeth	12
Large sprocket teeth	45

By referring to the torque curve shown in *Figure 7.2*, comment on the suitability of 1st and 2nd gear for accelerating off-the-line.

Solution

Total 1st gear ratio = 1.822 × 2.833 × 45/12 = 19.36

Required T_{engine} = 820/19.36 = **42.4 Nm**

Total 2nd gear ratio = 1.822 × 2.062 × 45/12 = 14.07

Required T_{engine} = 820/14.07 = **72.3 Nm**

Figure 7.4 shows these engine torque requirements compared to the engine torque output from *Figure 7.2*.

Figure 7.4
Torque requirements in first and second gears

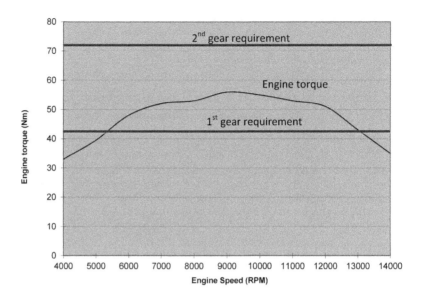

Comments:

1. It can be seen that, in this case, adequate torque is available in 1st gear between about 5500 rpm and 13 000 rpm. In this range the driver must be careful to avoid wheelspin.
2. At the start the driver must slip the clutch to keep the revs above 5500 until the car gains sufficient speed. In this case it entails travelling about 4 m for just less than a second before the clutch is fully released. If the driver releases the clutch too quickly, engine revs will drop and there is insufficient torque available. The car is likely to 'bog down'.
3. When sitting on the line the driver would probably hold the engine at about 8000 rpm where plenty of torque is available, however excessive wheelspin must be prevented by simultaneous control of the throttle and clutch.
4. It can be clearly seen that the engine cannot provide enough torque at any point to enable a successful start in 2nd gear. The driver would need to slip the clutch for about 20 m to avoid 'bogging down' when the clutch is fully engaged. This would generate considerable heat and wear of the clutch plates. In this case, after about 20 m, acceleration is governed by the power limit rather than grip.
5. It can be seen from the gear ratios given at the start of the example that there is a large gap between 1st and 2nd gears. This use of a 'short' first gear is common for road motorcycles and the racing driver would prefer a more even close ratio box. In this case the driver needs to hold the car in 1st gear up to the rev limit before changing gear.
6. With a chain-driven motorcycle engine, the designer does, of course, have the option of modifying all the gear ratios by changing the final drive sprocket ratio. This would lower or raise both lines in Figure 7.3 relative to the engine torque curve.

7.2.3 Top gear

The ratio of top gear depends upon the maximum speed that can be achieved on the track. We saw in *Figure 1.11* that cars have a theoretical maximum terminal speed when all the available power is used up in overcoming aerodynamic and other losses. In the unlikely event that there is a straight long enough to aproach this, the top gear ratio is set so that the top speed occurs at the peak power point – i.e. 12 000 rpm in the case of *Figure 7.2*. In general, however, the terminal velocity is not achievable and top gear is set so the car reaches its rpm limit just above the maximum speed that is expected for similar cars on a particular circuit. Hill-climb cars tend to have a lower top gear than circuit racers as they have less opportunity for high speeds.

7.2.4 Intermediate gears

Having established ratios for first and top gear it is now necessary to determine the distribution of the intermediate ratios. Gear ratios tend to become gradually closer together in the higher gears – a phenomenon known as **progression**. For a circuit which involves a lot of acceleration through the gears out of slow corners, a fairly linear progression may be appropriate, however if most of the racing takes place around a particular speed it may be better to close-up the gears so that the driver can keep the engine operating close to peak power for most of the lap.

Knowing the wheel rolling radius, R_r (m), and total gear ratios (i.e. engine r.p.m./wheel r.p.m.), it is an easy matter to evaluate car speed at specific increments of engine rpm for each gear:

For instance, at 1000 rpm

$$\text{Speed, (m/s)} = \frac{1000 \times 2\pi R_r / \text{total gear ratio}}{60}$$

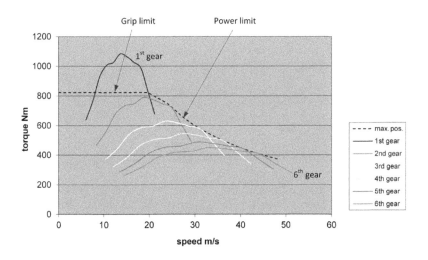

Figure 7.5
Torque at the wheels available in each gear

We also saw above how to determine the torque available at the wheels at specific increments of engine revs for each gear:

Wheel torque, $T_{wheels} = T_{engine} \times$ total gear ratio

The results can be combined and plotted to give *Figure 7.5*. This shows that the driver needs to adopt a specific gear at a given speed in order to maximise the torque available at the wheels for acceleration. At any speed the top-most curve indicates the best gear for wheel torque. It effectively tells the driver at what speed to change gear to maximise performance. It should be noted that:

1. Rather than memorising change speeds for each gear the driver can achieve exactly the same effect by driving around the peak power point (at 12 000 rpm in the case of *Figure 7.2*). This means changing up at say 12 700 rpm at which point the revs would drop to around 11 300 rpm.
2. The dashed line indicates the theoretical maximum torque available at the wheels taking into account traction grip and peak engine power. This is a similar shape to the traction force curve shown in *Figure 1.11*.
3. The individual gear curves touch the dashed line at the point where the engine reaches peak power. A car fitted with a **continuously variable transmission (CVT)** should be able to follow the dashed line, thus maximising available torque at the wheels. However, discrete gears produce 'valleys' between the peaks.
4. *Figure 7.5* reinforces the previous suggestion that the 'short' first gear, in this example, means that the driver must take the car close to the rpm limit before changing up to second. Also these ratios clearly close-up in the higher gears indicating that this particular car is good for racing at speeds around 42 m/s (150 km/h), where close to maximum torque is available in 4th, 5th and 6th gears.
5. The maximum speed in 6th gear in *Figure 7.5* is about 52 m/s (190 km/h). FSAE/Formula Student cars reach a maximum speed of around 35 m/s (125 km/h) in the acceleration event and hence it is common to ignore (or even remove) 5th and 6th gears.

7.3 Differentials

The differential distributes power from the gearbox to the driven wheels in such a way that, when cornering, the inner wheel can rotate more slowly than the outer wheel. The average speed of the two wheels is proportional to the input speed.

7.3.1 Differential types

The designer has several options:

- **The locked differential** (or no differential) – here the two driven wheels are locked together so that they are constrained to rotate at the same speed. This means that during the early stages of cornering the inner wheel will skid as it is forced to rotate faster than its movement over the ground requires. The outer wheel will skid as it is forced to rotate more slowly than required. The result is a reluctance to turn-in and initial understeer. However, at the limit, once the turn becomes established and lateral load transfer takes place, all of the slip will take place at the lightly loaded inner wheel. We saw from *Figure 5.6* that once a wheel starts slipping its level of grip reduces, hence more torque is applied at the outer wheel, resulting in a switch to oversteer. Considerable tyre wear is expected from the scuffing.
- **The open differential** – this is the type used on most road cars. The driven wheels are free to rotate at different speeds. It is a feature of this type of differential that torque is divided equally between the driven wheels. This is fine so long as both wheels maintain good grip, however if a wheel loses grip (and hence torque) through say a patch of ice, grass or gravel, the differential will react by spinning-up the slipping wheel and sending the same low torque to the gripping wheel. This can also occur as a result of lateral load transfer when accelerating a powerful car out of a corner. Allan Staniforth (*ref. 25*) says:

  > 'When the lesser loaded wheel finally loses grip and starts spinning the peculiarity of the open differential is that the power transfers itself to the spinning wheel, thus chucking it all away just when it is desperately needed to advance your progress or career.'

- **The limited-slip differential** – some form of limited-slip or **torque-biasing** differential is preferred on most racing cars if regulations permit. The driven wheels are partially locked together in such a way that they can rotate at different speeds when cornering, but the differential removes torque from the fastest spinning wheel and adds it to the slowest. This means that when cornering with good grip, the slow-moving inner wheel gets extra torque, producing an understeer effect. There are two main methods of achieving partial wheel locking. Firstly, by connecting the two output drive-shafts together with a spring-loaded clutch. Spring stiffness determines the slip threshold and hence the aggressiveness of the differential. The second approach is to use helical gears to develop friction under load. Such a system increases the friction, and hence the degree of locking between the wheels, as the input torque increases. Hence such systems are sometimes referred to as **torque sensing** differentials. Popular makes in racing are Quaife and Torsen.

7.3.2 Supporting the differential

With motorcycle-engined cars the differential is supported in such a way that its position can be adjusted to set the appropriate tension in the drive-chain. Several techniques have been adopted to achieve this including turnbuckles, shims and eccentric rotating bearing housings. Whatever system is used it must be borne in mind that the supports must resist the maximum tensile force in the chain and as such this is probably the highest stressed area of the car.

The maximum load in the chain can be determined in two ways:

1. from the maximum torque required to spin the wheels taking into account longitudinal load transfer during acceleration, or
2. from the maximum torque that can be generated by the engine multiplied by the total gear ratio in 1st gear.

With racing cars the second method generally gives the larger value and its adoption can perhaps be justified by considering the additional torque required to unstick a slick from hot tarmac. In addition, it is prudent to include a dynamic magnification factor of say 1.3 to account for sudden clutch release. The chain tension is then evaluated by dividing the torque by the radius of the large rear sprocket. This force is then applied as a load at the centre of the differential support bearings (torsion on the sprocket being carried by the drive shafts).

EXAMPLE 7.2

A FSAE/Formula Student car has wheel torque and gear ratios as given in *Example 7.1* and engine output as in *Figure 7.2*. The large rear sprocket has a diameter of 250 mm. The differential is supported as shown in *Figure 7.6*. Determine:

(1) the maximum chain tension, F_{chain},
(2) the differential bearing loads,
(3) the differential support reaction forces – H_1 to H_4, V_1 and V_2.

(1) From *Example 7.1*:

$$\text{Max. wheel torque} = 820 \text{ Nm}$$
$$\text{Torque from engine} = 55 \times 19.36$$
$$= 1065 \text{ Nm}$$
$$\text{Sprocket radius} = 0.25/2$$
$$= 0.125 \text{ m}$$
$$F_{chain} = 1.3 \times 1065/0.125$$
$$= 11\,076 \text{ N or } \mathbf{11.1 \text{ kN}}$$

Chapter 7 Rear wheel assembly and power transmission

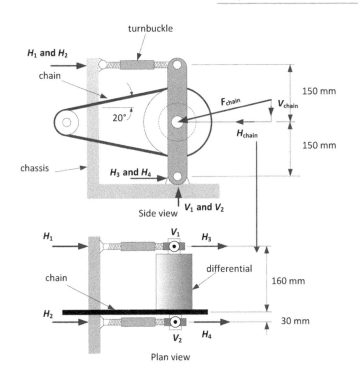

Figure 7.6
Forces on differential supports

(2) From *Figure 7.6*:

Bearing load at chain end = 11.1 × 160/190   = **9.3 kN**

(3) Resolving vertically:

$$V_2 = 9.3 \sin 20° = \mathbf{3.2\ kN}$$

Resolving horizontally and dividing by 2 from symmetry:

$$H_2 = H_4 = 9.3 \cos 20°/2$$
$$= \mathbf{4.4\ kN}$$

Bearing load at non-chain end = 11.1 × 30/190   = **1.8 kN**

Resolving vertically:

$$V_1 = 1.8 \sin 20° = \mathbf{0.6\ kN}$$

Resolving horizontally and dividing by 2 from symmetry:

$$H_1 = H_3 = 1.8 \cos 20°/2$$
$$= \mathbf{0.8\ kN}$$

7.4 Driveshafts

Driveshafts transmit torque from the differential to the wheels. They normally have splined ends to attach to universal joints. There are two commonly used forms of construction:

- Small-diameter solid or gun-drilled shafts made from very-high-quality alloy steel such as aerospace grade S155. Such steels require heat treatment to harden and temper the material after machining.
- Larger-diameter thin-walled hollow shafts which can be made from steel, aluminium, titanium or carbon fibre. These can be lighter but the greater volume can cause packaging problems. There is also the issue of welding or bonding the splined ends to the shafts.

Torsion generates **shear** in a circular shaft. Calculations should be performed to check that the maximum shear stress, τ_{max}, is less than the allowable value, τ_y, which can be assumed to be the material yield or proof stress, σ_y, divided by $\sqrt{3}$:

i.e. $\qquad \tau_y = \sigma_y/\sqrt{3} \qquad = 0.58\sigma_y$

The following standard formulae give the maximum **elastic** shear stresses in shafts.

For a solid circular shaft of radius R subjected to a torque T:

$$\tau_{max} = 2T/\pi R^3 \qquad [7.1]$$

For a gun-drilled shaft with a bore of r:

$$\tau_{max} = 2TR/\pi(R^4 - r^4) \qquad [7.2]$$

For a thin walled hollow shaft with wall thickness t, either use *equation [7.2]* (with $r = R - t$) or the following approximation:

$$\tau_{max} = T/2\pi R_{mean}^2 t \qquad [7.3]$$

where R_{mean} is the average of the inner and outer radii.

The most common point of driveshaft failure is at the interface between the shaft and the splined ends. Stress-raising corners at the root of the splines initiate torsion cracks. An essential design detail is to include a smooth transition into a reduced diameter neck at the ends of the splines – i.e. the diameter at the root of the splines should be greater than the shaft diameter as shown in *Figure 7.7 on page 188*.

EXAMPLE 7.3

A drive shaft is required to transmit a design torque of 600 Nm. Determine suitable diameters for the following two options:

(1) a solid shaft made from S155 steel with a proof stress of 1550 N/mm²,
(2) a hollow shaft made from 6082 T6 aluminium with a proof stress of 260 N/mm and a wall thickness of 2 mm.

(1) From *equation [7.1]*:

$$\tau_{max} = 2T/\pi R^3$$

$$R^3 = 2T/\pi \tau_{max} = (2 \times 600 \times 10^3)/(\pi \times 0.58 \times 1550)$$

$$= 424.9$$

$$R = 7.52 \text{ mm}$$

Solid diameter = say 16 mm

(2) From *equation [7.3]*:

$$\tau_{max} = T/2\pi R_{mean}^2 t$$

$$R_{mean}^2 = T/2\pi \tau_{max} t = (600 \times 10^3)/(2 \times \pi \times 0.58 \times 260 \times 2)$$

$$= 317$$

$$R_{mean} = 17.8 \text{ mm}$$

$$R_{outer} = 17.8 + 1 = 18.8 \text{ mm}$$

Outside diameter of hollow tube = say 38 mm

Comments:
1. *Substituting the hollow shaft result into the more precise* equation [7.2] *indicates a shear stress about 0.5% higher than that suggested by the approximate method.*
2. *The solid steel shaft weighs about 2.5 times the hollow aluminium alloy shaft.*

7.5 Universal joints

With an independent suspension, such as the double wishbone, the driveshafts must be fitted with some form of universal joint at each end to accommodate relative movement between the wheel and the differential. As well as angular movement, there must be a means of coping with a change in length of the driveshaft which is known as **plunge**. The traditional solution was to use a phased pair of **Cardan** or **Hook** joints combined with a splined sliding mechanism; however more recently some form of **constant velocity (CV) joint** is generally used. CV joints can accommodate both angular and longitudinal movement and two of the several different types available are:

Figure 7.7
Tripod CV joint

- **Rzeppa CV joints** which contain six balls that move in internal grooves. They can accommodate very large angle changes of around 50° which makes them suitable for the outer joints of front-wheel-drive road cars. They also allow around 10–20 mm of plunge movement.
- **Tripod CV joints** (*Figure 7.7*) are simpler and contain three rollers that move in a machined housing. They accommodate less angular rotation than Rzeppa joints (about 20°) but this is sufficient for rear wheel drive cars. They can also cope with up to 50 mm of plunge simply by extending the length of the housing. They are regarded as more efficient and can be lighter than Rzeppa joints and hence are popular for racing. If plunge is permitted at both ends of a driveshaft it is necessary to provide spring-loaded **snubbers** to centre the shaft and prevent lateral movement when

Figure 7.8
Rear upright assembly incorporating CV joint housing

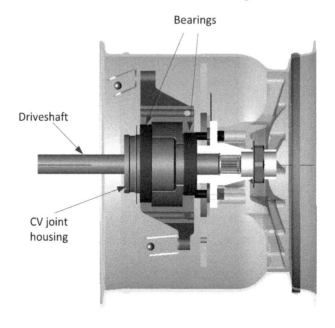

subjected to lateral g force on corners. An alternative is to restrict significant plunge movement to only one end of the shaft.

It is good practice to make the driveshafts as long as possible to reduce angle changes during suspension movement. One means of achieving this is to incorporate the outer CV joint housing into the axle which sits in the upright. *Figure 7.8* shows such a rear upright assembly. This approach also tends to reduce the number of parts but at the expense of larger uprights and bearings.

7.6 Rear wheel bearings

In *Chapter 6* it was suggested that, for the front wheels, a pair of either angular contact ball bearings or tapered roller bearings were to be used. These may also be appropriate for the rear wheels, however, if, as suggested above, the intention is to incorporate the CV joint housing into the upright, it will be found that these types are excessively large and heavy. The following types are appropriate:

- **Deep grooved ball bearings** are the commonest form of bearing and are available in a wide range of sizes and weights. They will resist both radial and axial forces and do not have to be used in opposing pairs like angular contact or tapered roller bearings.
- **Cylindrical roller bearings** support higher loads than deep grooved ball bearings of the same size. They are available with shoulders to resist axial loads although the axial load must be less than half the radial load.

As with the front wheel assembly, the bearings must be properly located in housings with shoulders to prevent lateral movement. The design procedure is identical to the front wheel bearings except that there is no thrust angle to increase the bearing spacing – see *Figure 7.9*. Also the load spectrum for the

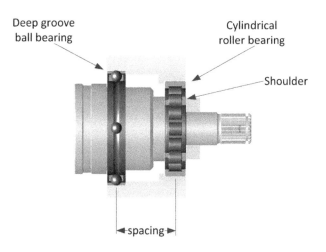

Figure 7.9
Rear bearing arrangement

Race car design

dynamic load check will contain the additional **maximum acceleration load case**. *Figure 3.14* shows how this involves a longitudinal force applied at wheel bearing level. This is treated in a similar manner to the maximum braking case as shown in *Example 6.2*.

7.7 Rear upright design

The rear upright is also subjected to the additional maximum acceleration load case as shown in *Figure 7.10*. As with the previous maximum braking case, taking moments about the centre of the inner bearing:

Figure 7.10
Loads on rear upright in maximum acceleration

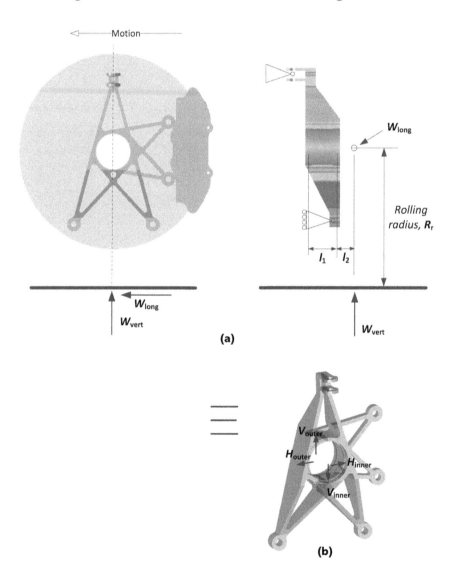

$$V_{outer} = \frac{W_{vert} \times (l_1 + l_2)}{l_1}$$

$$H_{outer} = \frac{W_{long} \times (l_1 + l_2)}{l_1}$$

Summing vertical forces:

$$V_{inner} = V_{outer} - W_{vert}$$

Summing horizontal forces:

$$H_{inner} = H_{outer} - W_{long}$$

7.8 Launch control, traction control and quick-shift

Most aftermarket engine control units (ECUs) contain the ability to assist the driver in certain situations although not all racing formulae permit their use.

Launch control removes the need for driver throttle control when accelerating off-the-line but still requires some clutch control. The system monitors the relative speed of rotation of the driven and undriven wheels and reduces engine power to control wheelspin. A speed sensor is required for an undriven front wheel together with the distance travelled in one revolution. The system is designed to operate in only the starting gear; hence the ECU can calculate the required speed of the driven wheels from knowledge of engine rpm, gear ratio and rolling circumference. The general procedure is typically:

1. While stationary at the start-line the driver engages launch control and fully depresses the accelerator pedal. Maximum engine rpm is controlled by the launch control system.
2. At the start, as the clutch is engaged, the system detects movement of the car and permits only the desired amount of wheel slip (say 10%) as the car accelerates.
3. Once the car reaches a set speed the system switches off and the engine is allowed to run freely.

It should be noted that:

- On the start line the engine rpm should be set to correspond to maximum engine torque (9000 in the case of *Figure 7.2*).
- It is important that the clutch is not released too quickly. The launch control system can reduce engine power but it cannot increase it. Once the clutch is fully engaged, the fixed relationship between road speed and

engine speed determines the engine output. Just as with a manual start, if the engine is rotating too slowly to provide the required power, the car will bog down.
- Different settings are required for wet and dry conditions.

Traction control limits wheelspin in all gears and generally requires additional wheel-speed sensors on both driven wheels as well as an undriven wheel. Knowledge of the current gear is also required and this can either be obtained from a gear position sensor or by estimation from engine rpm and car speed.

Quick-shift is a facility available for use with sequential gearboxes that enables the driver to make faster clutchless gear-changes without lifting off the throttle. It simply consists of a switch, usually connected to the gear-lever or paddle that sends a signal to the ECU when a gear-change is imminent. The ECU then momentarily unloads the transmission by cutting the engine.

SUMMARY OF KEY POINTS FROM CHAPTER 7

1. Engine power is proportional to torque × rpm.
2. The first gear ratio should be set so that, when the clutch is released, the wheel torque is sufficient to accelerate the car at the traction limit. If the gear is too high the car can 'bog down'.
3. The top gear ratio should be set so that the car reaches the desired maximum speed when close to the rpm limit.
4. Intermediate gears are generally **progressive**, i.e. the ratios get closer together in the higher gears. In order to maximise performance a driver should keep the engine as close as possible to the rpm at which maximum power occurs.
5. Some form of limited slip differential is desirable. With motorcycle-engined cars the forces required to restrain the differential can be very high.
6. Driveshafts can be solid or hollow and are designed to withstand torsion. They often have splined ends to suit **constant velocity joints**.
7. Deep grooved ball bearings and cylindrical roller bearings can be used if it is desired to accommodate the CV joint housing in the rear upright.

8 Brakes

> **LEARNING OUTCOMES**
>
> At the end of this chapter:
> - You will know the elements of a car braking system
> - You will understand the key objectives in brake system design
> - You will understand the importance of brake balance and how it is achieved
> - You will know how to size and specify the various brake components
> - You will learn how to evaluate the loads on the brake pedal and its assembly, and to ensure adequate robustness

8.1 Introduction

A highly effective braking system is vital for good lap times as it enables the driver to maintain high speeds for longer on the straights before **late braking** at a corner. They are also clearly a vital safety measure. *Figure 8.1* shows the main components of a typical racing car brake system. It is a hydraulic system which is pressurised by the driver applying a force to the brake pedal. Carroll Smith (*ref. 24*) says:

> *'The big payoff of a well sorted out braking system comes, not from any increase in braking power itself, but in the confidence, consistency and controllability that it provides to the driver.'*

The brake **discs** or **rotors** are traditionally made of cast iron which has been found to possess good frictional and thermal properties. It is used down to disc thicknesses of about 4 mm on motorcycles at which point it starts to become fragile. Thinner discs are made from steel but warping when hot becomes a problem with very thin discs. Powerful cars need powerful brakes and there is more energy to dissipate in the form of heat in the brakes. For this reason such cars use **ventilated discs** which consist of a double-wall disc with cooling vanes in-between. In cast iron, such discs are heavy and can form a significant proportion of the unsprung weight of the wheel assembly. To reduce weight, carbon composite and ceramic composite discs have been

Figure 8.1
Brake system components

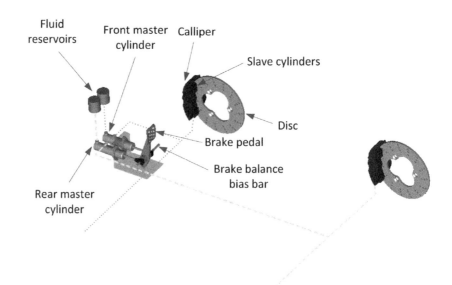

developed but they are very expensive. Aluminium brake discs are available and they work best with either a special coating and/or special brake pads. They are mainly suitable for short hill-climbs and sprints, as the relatively low melting point of aluminium is a concern with longer events where the discs may reach temperatures of 700°C (over 1000°C with carbon composite discs in F1). With temperatures like these it is not surprising that a common cause of brake failure is overheating. If this proves to be a problem the first step is to improve the airflow over the brake discs by adding **brake ducts**.

A key element of the braking system is the hydraulic brake fluid and it pays to use racing-quality fluid which has a higher safe operating temperature than standard fluid.

8.1.1 Basic functionality

The brake pedal is a lever which magnifies the applied force usually five or six times (*Figure 8.2*). The force is transmitted to the bearing on the adjustable brake bias bar which distributes it to the two master cylinders – one for the front brakes and one for the rear. If the force applied by the driver is F_{driver}:

Force applied to bias bar, $F_b = F_{driver} \times P_1/P_2$ [8.1]

If the threaded bias bar is turned, the total distance between the master cylinder shafts, $B_1 + B_2$, remains constant but the relative proportions of B_1 and B_2 change. From the reactions of a simply supported beam:

Force to front master cylinder, $F_f = F_b \times B_1/(B_1 + B_2)$ [8.2]

Force to rear master cylinder, $F_r = F_b - F_f$ [8.3]

Chapter 8 Brakes

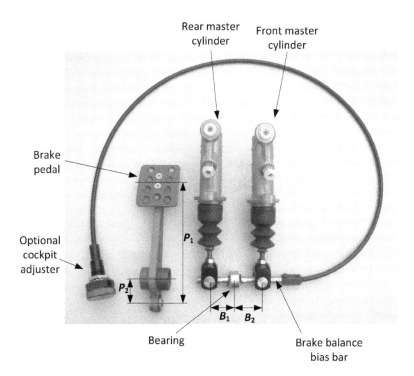

Figure 8.2
Brake pedal and balance bias bar

Each master cylinder contains a piston of specified diameter which pressurises the fluid in the system. If the front master cylinder, for example, has a piston of area A_{mf}:

Front fluid system pressure, $P_{bf} = F_f/A_{mf}$ [8.4]

Hydraulic hoses then transmit the pressure to **slave cylinders** which are contained within the brake calliper on each wheel. Slave cylinder pistons move to clamp *friction pads* onto the discs. There are typically between one and six slave cylinder pistons in each calliper. The callipers may contain slave cylinders on only one side of the disc; in which case either the disc or the calliper body must float so that the clamping force can be equalised on each side of the disc. More commonly, callipers are fixed and contain slave cylinders that push on both sides of the disc with equal force.

For the front slave cylinder with a piston area of A_{sf}:

Front calliper clamping force, $F_{cf} = P_{bf} \times A_{sf}$ [8.5]

In the above case the piston area, A_{sf}, is the total *on one side* of a disc.

If the centre of area of the friction pads is at a radius of r_b from the wheel centre and the coefficient of friction between the pad and the disc is μ_b, then for braking on both sides of the disc:

Front brake torque, $T_{bf} = 2 \times F_{cf} \times r_b \times \mu_b$ [8.6]

Values of μ_b are typically in the range 0.4 to 0.5 for steel or cast-iron discs.

8.2 Brake system design

8.2.1 Design objectives

1. The brakes in small racing cars rarely have servo-assistance and drivers should be able to lock all four wheels by applying reasonable pedal force – say no larger than that required to lift half their own body-weight or 375 N.
2. The brake system must be balanced. *Figure 5.6* shows that the braking force peaks at a slip ratio of about 10–15% and then reduces as the wheels start to lock. This means that for car stability during braking the front brakes should lock just before the rears. Theoretical design can get close to achieving this, but fine-tuning on the track is required, especially as road surface and weather conditions affect balance.
3. The hydraulic pressure in the system must not exceed safe levels for the brake components – manufacturers usually quote about 7 N/mm² (70 bar or 1000 psi).
4. The brake pedal and its assembly must be robust enough to resist a much higher force during panic braking in a potential accident. FSAE/Formula Student cars require a pedal force of 2000 N to be considered.
5. Brake hoses and other components should be routed and protected so that they are not damaged by heat or contact with the ground or moving parts.

8.2.2 Design procedure

Design of the braking system starts by assuming a coefficient of friction, μ, between the tyre and the road and calculating wheel loads as a result of longitudinal weight transfer as shown in *Example 1.4*. (Once vertical wheel loads have been calculated it is possible to refine the calculation by using a tyre model such as Pacejka or test data to more accurately estimate the coefficient of friction at the front and rear wheels and then repeat the load transfer and torque calculations.) The resulting wheel loads are multiplied by μ and the rolling radius of the tyre, R_r, to obtain the brake torques, T_{bf} and T_{br}. These calculated torques indicate the target brake balance between the front and rear wheels. Front/rear balance can be adjusted by any, or all, of the following means:

- different diameter brake discs,
- different sized pistons in the calliper slave cylinders,
- different system pressures by using different sized master cylinders for the front and rear,
- different system pressures by offsetting the brake bias balance bar.

Having established the target front and rear brake torques, calliper sizes and disc diameters can be provisionally selected. Again using the front brakes as an example:

From *equation [8.6]*
$$\text{Front calliper clamping force, } F_{cf} = T_{bf}/(2 \times r_b \times \mu_b) \quad [8.7]$$

From *equation [8.5)*
$$\text{Front fluid system pressure, } P_{bf} = F_{cf}/A_{sf} \quad [8.8]$$

The above calculation is repeated for the rear and the fluid system pressures, and P_{bf} and P_{br} compared in order to select provisional sizes for the master cylinders. The area of the pistons, A_{mf} and A_{mr}, should be roughly **inversely proportional** to the pressures although the range of standard master cylinder sizes is somewhat limited.

From *equation [8.4]*
$$\text{Force on front master cylinder, } F_f = P_{bf} \times A_{mf} \quad [8.9]$$
$$\text{Force on rear master cylinder, } F_r = P_{br} \times A_{mr} \quad [8.10]$$

From *equation [8.3]*
$$\text{Force applied to bias bar, } F_b = F_r + F_f \quad [8.11]$$

Assuming a maximum force input from the driver of say 375 N:

From *equation [8.1]*
$$\text{Minimum pedal ratio, } P_1/P_2 = F_b/375$$

If the spacing between the master cylinders is fixed, i.e. $B_1 + B_2$ in *Figure 8.2*, the initial trial position of the brake balance bias bar can be calculated:

From *equation [8.2]*
$$B_1 = (B_1 + B_2) \times F_f/F_b \quad [8.12]$$

EXAMPLE 8.1

For the car shown in *Figure 8.3* design a braking system assuming the following data:

Tyre/road friction coefficient, μ = 1.7
Pad/disc friction coefficient, μ_b = 0.5
Wheel rolling radius, r_r = 270 mm
Radius to centre of all pads, r_b = 100 mm throughout.

(Consideration could be given to using larger discs on the front; however, in this example the main means of setting the required bias towards the front will be by sizing the master cylinders and using callipers with different slave cylinder areas at the front and the rear. Fine adjustment is then by means of the bias balance bar.)

Race car design

Figure 8.3
Brake system design

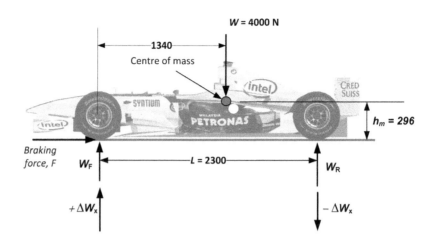

Front calliper piston area, A_{sf} = 1587 mm²
Rear calliper piston area, A_{sr} = 1019 mm²
Spacing of master cylinders = 65 mm

Master cylinders to be selected from the following:

Imperial size (inch)	Piston area (mm²)
0.625	198
0.700	248
0.750	285
0.813	335

Solution

Static axle loads

$W_R = 4000 \times 1340/2300 = 2330$ N

$W_F = 4000 - 2330 = 1670$ N

Braking force $F = 4000 \times 1.7 = 6800$ N

From *equation [1.5]*

Longitudinal weight transfer, $\Delta W_x = \pm \dfrac{F h_m}{L} = \pm \dfrac{6800 \times 296}{2300}$

$= \pm 875$ N

Front wheel loads W_{FL} and $W_{FR} = \dfrac{1670 + 875}{2} = 1272$ N (55%)

Rear wheel loads W_{RL} and $W_{RR} = \dfrac{2330 - 875}{2} = 728$ N (45%)

Chapter 8 **Brakes**

Front wheel brake force	= 1272 × 1.7	= 2162 N
Rear wheel brake force	= 728 × 1.7	= 1238 N
Front wheel brake torque, T_{bf}	= 2162 × r_r	= 2162 × 270
	= 584 000 Nmm	
Rear wheel brake torque, T_{br}	= 1238 × r_r	= 1238 × 270
	= 334 000 Nmm	

From *equation [8.7]*
Front calliper clamping force, $F_{cf} = T_{bf}/(2 \times r_b \times \mu_b)$
$= 584\,000/(2 \times 100 \times 0.5)$
$= 5840$ N

From *equation [8.8]*
Front fluid system pressure, $P_{bf} = F_{cf}/A_{sf}$ = 5840/1587
$= 3.66$ N/mm² < 7.00 N/mm² ✓

From *equation [8.7]*
Rear calliper clamping force, $F_{cr} = T_{br}/(2 \times r_b \times \mu_b)$
$= 334\,000/(2 \times 100 \times 0.5)$
$= 3340$ N

From *equation [8.8]*
Rear fluid system pressure, $P_{br} = F_{cr}/A_{sr}$ = 3340/1019
$= 3.28$ N/mm² < 7.00 N/mm² ✓

Looking at the relative magnitudes of the front and rear system pressures, try a 0.700 inch master cylinder for the front and a 0.75 inch for the rear.

From *equation [8.9]*
Force on front master cylinder, $F_f = P_{bf} \times A_{mf}$ = 3.66 × 248
$= 908$ N

From *equation [8.9]*
Force on rear master cylinder, $F_r = P_{br} \times A_{mr}$ = 3.28 × 285
$= 935$ N

From *equation [8.11]*
Force applied to bias bar, $F_b = F_r + F_f$ = 908 + 935
$= 1843$ N

Assuming a maximum force input from the driver of say 375 N:

Minimum pedal ratio, P_1/P_2 = 1843/375 = 4.9

(an actual ratio of between 5 and 6 is typical)

If the spacing between the master cylinders, $B_1 + B_2$, is 65 mm

From *equation [8.12]*

$$B_1 = (B_1 + B_2) \times F_f/F_b = 65 \times 908/1843$$
$$= 32.0 \text{ mm}$$

i.e. this is an initial offset from the central position of only 0.5 mm to the left.

Comment:
Final adjustment is made as a result of testing on the circuit. Figure 5.6 indicates that braking force reaches a peak value which then reduces as the tyre starts to skid. This means that when one end of the car locks-up there is a drop in effective braking at that end. Somewhat surprisingly therefore, we want the front brakes to lock fractionally before the rears. This keeps the car stable and prevents the back-end from overtaking the front!

8.2.3 Effect of downforce

Example 8.1 is appropriate for zero downforce cars or high downforce cars at slow speeds when aerodynamic forces are low. For high speed braking of high downforce cars, the calculations should be repeated using the resulting higher wheel loads as shown in *Example 1.5*. Even taking into account a reduced coefficient of friction at the tyre/road interface from tyre sensitivity, the braking torques on all the wheels will be significantly higher. This will clearly increase hydraulic system pressures and the resulting pedal force. Unfortunately it will probably also affect the optimum front/rear balance of the brakes. This means that the position of the brake bias balance bar may have to be set at a compromise between fast and slow braking.

SUMMARY OF KEY POINTS FROM CHAPTER 8
1. Effective brakes are important for both safety and performance.
2. The system is hydraulic and consists of two master cylinders connected to the brake pedal and a series of slave cylinders at each disc brake calliper.
3. It is vital that the components are selected and sized to enable a good balance between the front and rear brakes.

9 Aerodynamics

LEARNING OUTCOMES

At the end of this chapter:
- You will know the major elements of a car aerodynamic package, particularly those elements that generate downforce
- You will be aware of those elements of fluid mechanics theory that are relevant to racing car aerodynamics
- You will understand how aerofoil wings work and how a particular type and size can be specified
- You will understand the principles behind the creation of downforce from the floor-pan
- You will learn how to carry out preliminary calculations to specify a balanced aerodynamic package

9.1 Introduction

Aerodynamics is a large and complex subject. The focus here is on those aspects of the subject that are relevant to the development of **downforce** on a racing car, as we saw in *Chapter 1* that the presence of downforce can increase braking and cornering performance greatly. Modern F1 teams devote significant resources to the optimisation of their **aero package** and it is generally regarded as the major factor that distinguishes the top teams from the rest. Simon McBeath (*ref. 14*) says:

> 'There is no other single aspect of competition car technology that has had such a big influence on performance as the exploitation of downforce.'

Figure 9.1 shows some of the devices currently used as part of an aero package. The interaction between devices is important so that the maximum benefit can be extracted from the air as it moves over the length of the car. The overall objective is to maximise downforce without creating too much drag. The main elements are the front and rear wings and the underside of the floor-pan which each contribute roughly a third of the downforce. As we shall see, aero forces increase in proportion to the square of the car's velocity and it is important that the downforce is *balanced* so that the understeer/oversteer characteristics of the car remain acceptable at different speeds.

Race car design

Figure 9.1
Aerodynamic devices (reproduced with kind permission of Caterham F1 Team)

9.2 Fluid mechanics principles

9.2.1 The properties of air

Air is a gaseous fluid mixture consisting of 78% nitrogen and 21% oxygen. Air is compressible and we know from the **gas laws** that for a given mass of air:

$$\frac{P_1 V_1}{T_1} = \text{constant} = \frac{P_2 V_2}{T_2} \qquad [9.1]$$

where
P_i = absolute pressure
V_i = air volume
T_i = absolute temperature.

We also know that the density, ρ, of a given mass of air is inversely proportional to its volume so an alternative to *equation [9.1]* is:

$$\frac{P_1}{T_1 \rho_1} = \text{constant} = \frac{P_2}{T_2 \rho_2} \qquad [9.2]$$

where ρ_n = air density.

Air density is important because it determines the magnitude of aerodynamic forces such as drag and downforce. (It is also important for the power output of internal combustion engines which depend upon the mass of oxygen that can be forced into the cylinders.) A value for air density is given in *Table 9.1*. It can be seen from *equation [9.2]* that density increases in proportion to absolute pressure and is inversely proportional to absolute temperature.

Table 9.1 Air data

Air data at 20°C and at sea level	
Atmospheric pressure, P_{sl}	101 325 N/m²
Density, ρ	1.204 kg/m³
Viscosity, μ	1.8 × 10⁻⁵ Pa sec

Chapter 9 Aerodynamics

A value for atmospheric pressure at sea level, P_{sl}, is given in *Table 9.1* but it reduces exponentially with elevation above sea level according to:

Atmospheric pressure at h metres above sea level $\approx P_{sl} \times e^{-(h/7000)}$ [9.3]

Air density is normally assumed constant for any particular situation; however, as shown in *Example 9.1* it can vary considerably between circuits.

Another important property of all fluids such as air is **viscosity**, μ. This is a measure of the force required to cause relative movement of one layer of air over another, and is to do with the interaction of moving gas molecules. It is clearly more difficult to stir a viscous fluid such as treacle than a less viscous fluid such as water. Viscosity also makes it harder to run in water than in air. Viscous forces are also time-dependent as we saw in relation to viscous dampers in *Chapter 4*. It requires a larger force to stir treacle quickly than it does to stir it slowly. The units of viscosity thus involve time, and the value for viscosity for air at 20°C at sea level is given in *Table 9.1*. This figure is largely independent of atmospheric pressure but increases slightly with temperature. Viscosity is relevant to racing cars because it is responsible for creating the **boundary layer** as described in *section 9.2.5*.

EXAMPLE 9.1

Estimate the percentage reduction in engine power and aerodynamic forces at the Brazilian GP in Interlagos (800 m elevation and 40°C) compared to the Monaco GP (sea level and 20°C).

Equation [9.3]

$$\begin{aligned}
\text{Atmospheric pressure at Interlagos} &= P_{sl} \times e^{-(h/7000)} \\
&= 101\,325 \times e^{-(800/7000)} \\
&= 101\,325 \times 0.892 \\
&= 90\,382 \text{ N/mm}^2
\end{aligned}$$

$$\begin{aligned}
\text{Absolute temperature at Monaco} &= 273° + 20° &= 293° \\
\text{Absolute temperature at Interlagos} &= 273° + 40° &= 313°
\end{aligned}$$

Equation [9.2]

$$\frac{P_1}{T_1 \rho_1} = \frac{P_2}{T_2 \rho_2} \quad [9.2]$$

$$\frac{101\,325}{293 \times 1.204} = \frac{90\,382}{313 \times \rho_2}$$

Density of air at Interlagos, ρ_2 = 1.005 kg/m³
(1.204 kg/m³ at sea level and 20°C)

% reduction = 100 × (1.204 − 1.005)/1.205

= **16.5%**

9.2.2 Laminar and turbulent flow

The flow of air can be in one of two states – **laminar** or **turbulent**, as shown in *Figure 9.2*. Air flow can be represented graphically in the form of streamlines which trace the movement of air particles. When the streamlines are ordered and parallel to the average direction of flow, the flow is said to be **laminar**. Usually after interaction with an object the flow can become disordered or chaotic which is known as **turbulent flow**. The transition from laminar to turbulent flow is often initiated by a build-up of viscous forces converting the kinetic energy of the fluid particles into heat. In general, most racing car aerodynamic devices, such as wings, perform much better in laminar flow and hence a key aim is to preserve laminar flow as the air moves towards the rear wing and diffuser. *Figure 9.2* also shows how cars tend to produce a turbulent **wake** and this explains why a car following too closely behind another has the effectiveness of its aerodynamic devices considerably reduced.

It should be noted that in *Figure 9.2* the car is depicted as stationary and the air flow moving. This is the situation during wind tunnel testing; however in reality of course the opposite situation occurs.

Figure 9.2 Laminar and turbulent flow

9.2.3 The Bernoulli equation

The Bernoulli equation gives good insight into many of the aerodynamic phenomena relevant to racing cars. The equation applies to non-viscous, non-compressible laminar fluid flow and assumes that the total energy of a particle along a streamline remains constant. This energy can be in one of three forms:

- kinetic energy from the mass and velocity of the fluid = $\frac{1}{2}mv^2$,
- pressure energy = PV,
- potential energy from the mass and elevation of the fluid = mh,

where v is velocity of flow and V is volume.

Thus $\quad \frac{1}{2}mv^2 + PV + mh = \text{constant}$

If the density of the fluid is ρ, then $V = m/\rho$. Substituting into the second term:

$$\tfrac{1}{2}mv^2 + Pm/\rho + mh = \text{constant}$$

For a fluid gas such as air the potential energy term is negligible and can be ignored. Also divide through by m and multiply by ρ:

$$\tfrac{1}{2}\rho v^2 + P = \text{constant}$$

Both remaining terms now have units of pressure where P is the **static pressure** and $\tfrac{1}{2}\rho v^2$, rather confusingly, is the **dynamic pressure**. (The dynamic pressure is the additional static pressure that would result if all the kinetic energy was converted into pressure energy.)

A useful form of the Bernoulli equation compares two situations at different points along a streamline :

$$\tfrac{1}{2}\rho v_1^2 + P_1 = \tfrac{1}{2}\rho v_2^2 + P_2 \qquad [9.4]$$

Here it can be seen that if the velocity of flow increases from v_1 to v_2 the static pressure P_2 must reduce compared to P_1.

EXAMPLE 9.2

The floor-pan of a racing car is designed to increase the speed of air flow beneath the car by a factor of 1.5 over an area of 1.3 square metres. Estimate the resulting downforce at 50 m/s.

Density of air $\rho = 1.204$ kg/m^3.

At a point just in front of the car the pressure is atmospheric:

$$v_1 = 50 \text{ m/s}, P_1 = 101\,325 \text{ N/m}^2$$

Under the car:

$$v_2 = 1.5 \times v_1 = 1.5 \times 50 = 75 \text{ m/s}; P_2 = ?$$

From *equation [9.4]*

$$\tfrac{1}{2}\rho v_1^2 + P_1 = \tfrac{1}{2}\rho v_2^2 + P_2$$

$$\tfrac{1}{2} \times 1.204 \times 50^2 + 101\,325 = \tfrac{1}{2} \times 1.204 \times 75^2 + P_2$$

$$P_2 = 99\,444 \text{ N/m}^2$$

$$\text{Downforce pressure} = 101\,325 - 99\,444$$

$$= 1881 \text{ N/m}^2$$

$$\text{Downforce from floor-pan} = 1881 \times 1.3$$

$$= \mathbf{2445 \text{ N}}$$

Comment:
*The above value of 2.445 kN is significant and demonstrates the potential of the floor-pan for generating downforce; however the difficulties of generating and maintaining negative pressure under the floor should not be underestimated. This is covered in more detail later. In particular, the **boundary layer** covered in section 9.2.5 has a significant impact.*

9.2.4 Reynold's number

Aerodynamic phenomena such as downforce and drag are often investigated by the use of scale models in wind tunnels. In order to ensure that the measured results from such tests are appropriate to the full-scale car, it is usual to ensure that the model tests have the same **Reynold's number** as the full-scale car. This approach is known as **dynamic similitude**.

$$\text{Reynold's number}, Re = \frac{\rho v L}{\mu} \quad [9.5]$$

where
- ρ = air density
- μ = air viscosity
- v = air velocity
- L = typical dimension, such as length of car.

Reynold's number is dimensionless. It can be seen that a third scale model would have to be tested with three times the air velocity to keep the number the same.

EXAMPLE 9.3
Determine a typical value of Reynold's number for a racing car 3 m long at 50 m/s.

Equation [9.5]
$$Re = \frac{\rho v L}{\mu}$$

Obtaining ρ and μ from *Table 9.1*
$$= \frac{1.204 \times 50 \times 3}{1.8 \times 10^{-5}}$$

$$= 10 \times 10^6$$

Comment:
The above Reynold's number is appropriate for the whole car bodywork or floor-pan but for say a single wing element, the reduced dimension of L results in a value of about 10% of this.

9.2.5 The boundary layer

It is a fact that when a fluid flows against a solid boundary, such as the body of a car, the velocity of the fluid at the actual interface is zero. The fluid effectively sticks to the solid at the contact surface. (There does not appear to be an easy and convincing explanation for this phenomenon but it is undoubtedly true even for highly polished surfaces!) Viscosity means that air close to the boundary is slowed down but it gradually increases to the free stream velocity at some distance away from the boundary. The thickness of the boundary layer is normally defined as the distance from the surface to a point at which the velocity has reached 99% of the free-stream velocity. When the air-stream first encounters a solid surface the boundary layer is laminar but, as the layer gradually loses energy, there is a transition to a thicker turbulent layer which lies above a thin viscous sublayer – *Figure 9.3*.

The presence of the boundary layer causes drag on the surface. It effectively makes a solid object appear slightly bigger than it actually is. The size of the boundary layer can be estimated by using the local Reynold's number, Re_x, which is obtained by substituting the distance x from the start of the boundary layer for the length term – see *Figure 9.3*.

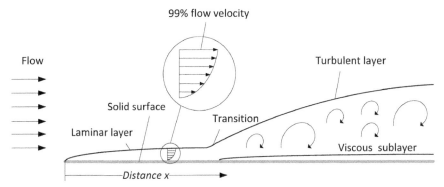

Figure 9.3
The boundary layer

$$\text{Local Reynold's number,} \quad Re_x = \frac{\rho v x}{\mu} \qquad [9.6]$$

The distance from the start of the boundary layer to the transition point depends upon surface roughness and the degree of turbulence in the main flow but has generally occurred before Re_x reaches 2×10^6. For the racing car at 50 m/s (*Example 9.3*), this implies the boundary layer is turbulent after about 20% of the car length, i.e. 600 mm from the front.

The thickness of the boundary layer can be estimated from the **Blasius solution**:

$$\text{Laminar boundary layer thickness} = \frac{4.91x}{\sqrt{Re_x}} \qquad [9.7]$$

$$\text{Turbulent boundary layer thickness} = \frac{0.382x}{Re_x^{1/5}} \qquad [9.8]$$

EXAMPLE 9.4

For a car travelling at 50 m/s estimate:
(a) the thickness of the laminar boundary layer 500 mm from the front,
(b) the thickness of the turbulent boundary layer 3000 mm from the front.

(a) Local Reynold's number, $Re_x = \dfrac{\rho v x}{\mu} = \dfrac{1.204 \times 50 \times 0.5}{1.8 \times 10^{-5}} = 1.67 \times 10^6$

Laminar boundary layer thickness $= \dfrac{4.91x}{\sqrt{Re_x}} = \dfrac{4.91 \times 0.5}{\sqrt{1.67 \times 10^6}} = 0.0019$ m

$= 1.9$ mm

(b) Local Reynold's number, $Re_x = \dfrac{\rho v x}{\mu} = \dfrac{1.204 \times 50 \times 3}{1.8 \times 10^{-5}} = 10 \times 10^6$

Turbulent boundary layer thickness $= \dfrac{0.382x}{Re_x^{1/5}} = \dfrac{0.382 \times 3}{(10 \times 10^6)^{1/5}} = 0.046$ m

$= 46$ mm

9.2.6 Drag

Aerodynamic drag on a vehicle is a force which acts in the direction opposite to the direction of motion. It therefore consumes engine power and hence impedes acceleration and reduces top speed. We have already noted that the boundary layer is a source of drag. This is known as **viscous** or **skin-friction drag**; however this is usually small compared to **form** or **pressure drag** which derives from the frontal area and shape of the vehicle.

Anyone who has put their hand out of a car window at speed or tried to carry a large sheet of plywood on a windy day knows that air flow can produce significant forces. We saw from Bernoulli's equation in *section 9.2.3* that dynamic pressure = ½ρv^2. If a simple rectangular sheet of plywood, of area A, was placed perpendicular to the direction of air flow, the velocity immediately in front of the sheet is zero and all of the dynamic pressure is converted into static pressure:

$$\text{Force} = \text{pressure} \times \text{area}$$

Force on front face pushing plywood back, from *equation [9.4]*

$$= \tfrac{1}{2}\rho v^2 A$$

Also turbulent vortices are released from the edges of the plywood causing negative pressure or suction on parts of the rear face of the plywood. This adds to the force pushing the plywood back and increases the total force on the sheet by about 20%:

$$\text{Total drag force} = C_D \times \tfrac{1}{2}\rho v^2 A \qquad [9.9]$$

where C_D is the dimensionless **drag coefficient** which has a value of about 1.2 in this case. The total drag force or **form drag** is the force that would be measured in a wind tunnel and is the sum of both skin friction and pressure drag. Clearly the plywood sheet described here is a particularly **bluff** shape and **streamlining** can reduce the value of drag coefficient considerably. At the other extreme is the streamlined **fusiform** shape of a dolphin where a drag coefficient of less than 0.01 is reported. *Table 9.2* provides some typical drag coefficients for various objects. Note the relatively high value for racing cars with wings where the generation of downforce is regarded as more important than minimising drag.

Table 9.2 Drag coefficients

Shape	Drag coefficient C_D
Rectangular sheet	1.2
Racing car (open wheel) with no wings	0.6
Racing car with wings	0.7–1.2
Typical modern road car	0.35
Modern Eco-car	0.25
Aircraft	0.012
Dolphin	<0.01

EXAMPLE 9.5

Calculate the power required to overcome aerodynamic drag on a racing car travelling at 60 m/s if it has a frontal area of 1.2 m² and a drag coefficient, C_D, of 0.8.

Equation [9.9] Drag force, F_D = ½ $C_D \rho v^2 A$ = ½ × 0.8 × 1.204 × 60² × 1.2

= 2081 N

Power used = $F_D \times v$ = 2081 × 60/10³

= 125 kW (168 bhp)

Comment:
Note that power consumption is proportional to the cube of the velocity.

9.2.7 Momentum

The concept of **momentum** is rarely discussed in terms of racing car aerodynamics; however it provides a powerful means of explaining the working of many aero devices – particularly wings. Because air has *mass*, Newton tells us that changing its direction of flow requires an external force. Also the force on the aerodynamic device is equal and opposite to the force on the air. These forces are required to generate a **rate of change in momentum** of the air.

Momentum is a vector quantity, i.e. it has both magnitude and direction.

$$\text{Momentum} = \text{mass} \times \text{velocity}$$

However because we are concerned with the *rate of change* of momentum we replace mass by **mass flow rate**. For a jet of fluid of cross-sectional area A, density ρ, and velocity v:

$$\text{Mass flow rate} = A\rho v \text{ kg/sec}$$

$$\text{Momentum rate} = A\rho v^2 \text{ kgm/sec}^2$$

Figure 9.4 shows a jet of air passing over a curved surface on a vehicle. We will consider the rate of change of momentum in the x and y directions between the two planes a–a and b–b. We will ignore the effect of skin-friction on the surface and assume that the velocity of flow at b–b is the same as that at a–a.

In the x-direction – force on fluid:

$$\text{Rate of change of } x \text{ momentum} = A\rho v^2 \cos\theta - A\rho v^2$$

$$= \boldsymbol{A\rho v^2(\cos\theta - 1) \text{ N}} \qquad [9.10]$$

In the y-direction – force on fluid:

$$\text{Rate of change of } y \text{ momentum} = -A\rho v^2 \sin\theta - 0$$

$$= \boldsymbol{-A\rho v^2(\sin\theta) \text{ N}} \qquad [9.11]$$

The forces on the curved surface are equal and opposite to the above where the y-direction force is downforce and the x-direction force is drag.

The two forces could be combined as vectors to find the resultant force:

$$\text{Resultant force} = \sqrt{((A\rho v^2(\cos\theta-1))^2 + (-A\rho v^2(\sin\theta))^2)} \text{ N}$$

With a concave surface, as shown in *Figure 9.4*, the air flow produces positive pressures on the top of the device. If these pressures were integrated over the entire surface area the force would be the same as the resultant calculated above.

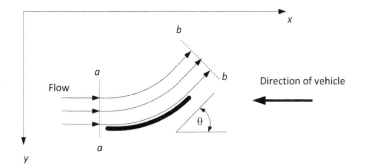

Figure 9.4
Change in momentum over curved surface

EXAMPLE 9.6

A curved surface on a car is assumed to deflect a jet of air whose cross-sectional dimensions are 1.2 m wide × 0.5 m high. If the car is travelling at 50 m/s compare the downforce and drag generated when the surface deflects the flow (a) at 8° and (b) at 16° from the horizontal.

$$\text{Momentum rate} = A\rho v^2 = 1.2 \times 0.5 \times 1.204 \times 50^2$$
$$= 1806 \text{ kgm/sec}^2$$

(a)   8° deflection

Equation [9.10]   Drag $= -A\rho v^2(\cos \theta - 1) = -1806(\cos 8° - 1)$
$$= 17.6 \text{ N}$$

Equation [9.11]   Downforce $= A\rho v^2(\sin \theta) = 1806(\sin 8°)$
$$= 251 \text{ N}$$

i.e. drag is about 7% of downforce.

(b)   16° deflection

Equation [9.10]   Drag $= -A\rho v^2(\cos \theta - 1) = -1806(\cos 16° - 1)$
$$= 70.0 \text{ N}$$

Equation [9.11]   Downforce $= A\rho v^2(\sin \theta) = 1806(\sin 16°)$
$$= 498 \text{ N}$$

i.e. drag is about 14% of downforce.

Comments:

1. It can be seen that as the deflected angle increases, the percentage of drag compared to downforce increases. This could be interpreted as less efficient, however the usual approach in racing car design is to decide on a **drag budget**, i.e. how much power can be devoted to overcoming drag and then to maximise the downforce.

2. *The size of the air jet used in the above example is clearly somewhat arbitrary. In the case of a jet of water striking a Pelton Wheel, say, it is easy to determine accurately the mass flow rate of the water jet and hence calculate the forces. In the case of air flowing in air this is much more difficult, and for this reason the performance of aero-devices is generally determined either by computer CFD (Computational Fluid Dynamics) modelling, wind tunnel testing of a scale model or, best of all, through full-scale testing of an instrumented car with data-logging on the circuit.*

9.3 Wings

The development of the **wing** was obviously initiated in the aircraft industry where the objective is to create lift. In motorsport, wings are inverted to generate **downforce**. The use of a simple inclined sheet of plywood or aluminium would create some downforce by deflecting the air and changing its momentum as described above; however it would be very inefficient compared to the use of a carefully designed **aerofoil** section. *Figure 9.5* shows a typical aerofoil together with the terminology used to describe it:

- The **chord** is a straight line joining the front of the **leading edge** to the back of the **trailing edge**.
- The inclination of the chord line relative to the incoming flow (assumed to be horizontal in *Figure 9.5*) is known as the **angle of attack**.
- The **thickness** is at the thickest point of the wing and it is often expressed as a percentage of the chord length.
- The **camber line** passes through the centre of the wing and shows the curvature of the wing cross-section. The maximum distance from the camber line to the chord line may also be expressed as a percentage of the chord length.

Figure 9.5
Aerofoil wing definitions

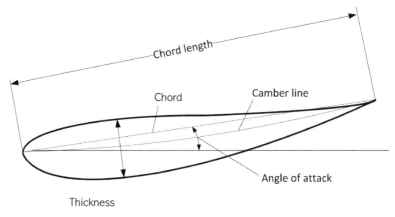

- The width of the wing (into the paper in *Figure 9.5*) is the **span**, and span/chord length is the **aspect ratio** of the wing.

9.3.1 The single-element wing

The 'magic' of wings is largely down to the fact that, because of viscosity and the resulting boundary layer, air flow sticks to the underside of the wing as well as the top. *Figure 9.6a* shows the streamlines as they pass the wing and they are said to be **attached**. Also shown is the pressure on the wing surfaces. It can be seen that, compared to atmospheric pressure, there is positive pressure on the top surface and negative pressure on the bottom – both of which contribute to downforce. In fact, the negative suction pressure on the underside generally contributes more to downforce than the positive pressure on the top. The pressure changes can be predicted to some extent by the Bernoulli equation as the airflow over the top of the wing slows down and that on the underside speeds up. It can also be seen from the streamlines that the wing

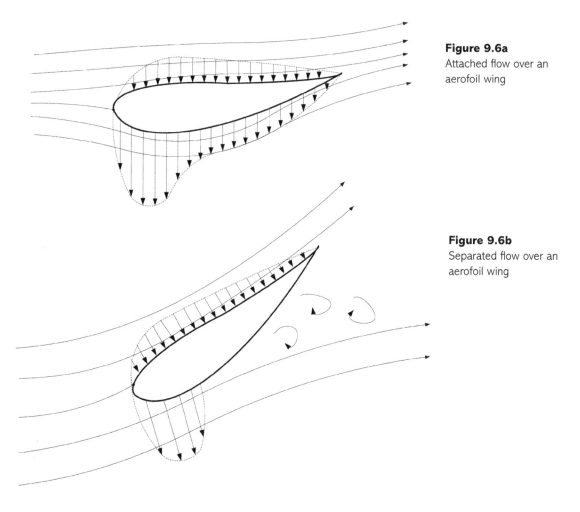

Figure 9.6a
Attached flow over an aerofoil wing

Figure 9.6b
Separated flow over an aerofoil wing

causes a change in the direction of the flow, resulting in forces to change the rate of momentum. In fact, it can be seen that the highest pressure on the underside of the wing occurs at the point where the curvature, and hence change of direction of air flow, is maximum. 'Bernoulli' and 'momentum' provide alternative explanations of the same phenomena.

Even at zero angle of attack, a cambered aerofoil can generate some downforce; however both downforce and drag increase as the angle of attack is increased – but only up to a certain point! When the angle of attack reaches a critical value, the flow **detaches** or **separates** from the bottom wing surface. Separation is a result of flow in the boundary layer losing energy from skin friction and being expected to move into a region of relatively higher pressure. It can be seen from *Figure 9.6b* that the area over which the suction pressure on the underside of the wing acts is greatly reduced, resulting in a fall in downforce. The wing can be said to have **stalled**. The positive pressure on the top of the wing is largely unaffected. Single-element wings typically reach peak downforce at an angle of attack of about 12°.

9.3.2 Multiple-element wings

Figure 9.7 shows a dual-element wing where a **flap** has been added to the **main-plane**. A small converging gap of about 12 mm between the two elements allows flow from the positive pressure area on top of the main-plane to accelerate through the gap and re-energise the negative pressure area on the underside of the flap. Such an arrangement enables the combined angle of attack to be almost doubled to about 20° before separation. This results in significant gains in downforce. Gains can be achieved by adding more elements but some formulae such as F1 limit the number of wing elements to two.

Figure 9.7
Dual-element wing

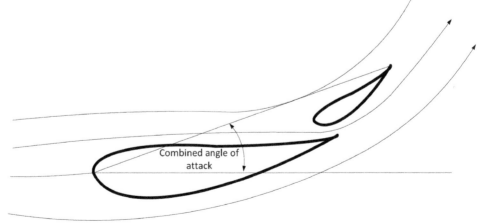

9.3.3 Wing selection

Many hundreds of wing profiles have been developed and published (*ref. 28*) and so it is possible to adopt an existing shape. This is particularly useful if you can find data from tests, ideally conducted at a similar Reynolds number, that reveal **lift coefficients** (C_L) and the **drag coefficients** (C_D) for your selected profile at different angles of attack. Of course, for cars the lift coefficient is actually a **downforce coefficient** as the wing is inverted.

For a wing:

$$\text{Downforce} = C_L \times \tfrac{1}{2}\rho v^2 A \qquad [9.12]$$

$$\text{Drag force} = C_D \times \tfrac{1}{2}\rho v^2 A \qquad [9.13]$$

Although *equation [9.13]* is identical to *equation [9.9]* it is a convention that, for wings, the area, A, is taken as the plan area (i.e. span × chord) and *not* the frontal area as previously discussed. *Table 9.3* gives approximate maximum values of C_L and C_D for wings with a different number of elements. Obviously a wing does not have to be operated at the maximum angle of attack shown in the table.

Table 9.3 Typical lift and drag coefficients

Number of elements	Angle of attack	Lift coefficient C_L	Drag coefficient C_D
1	12°	1.2	0.3
2	20°	2.2	0.7
3	26°	3.0	1.2

As an alternative to adopting an existing profile, designers may choose to define their own. In this case, lift and drag coefficients can be estimated from simulation in a computer CFD package.

Other general wing design issues are:

- End plates improve the efficiency of wings by reducing the leakage of downforce pressures at the ends.
- Adding increased camber to a wing profile is somewhat similar to increasing the angle of attack.
- The thickness of the profile is generally around 15–20% of the chord length.
- The wing span should be made as large as regulations permit to maximise efficiency.
- The addition of small 'Gurney flaps' to the upper trailing edge of wings can increase downforce and permit a higher angle of attack before separation occurs. The flaps protrude about the thickness of the boundary layer into the flow – see *Figure 9.8*.

Race car design

Figure 9.8
The Gurney flap

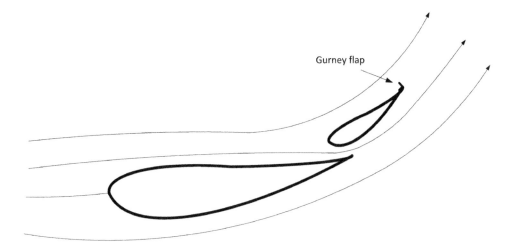

- Ideally the angle of attack for all wings should be adjustable to enable tuning of the aerodynamic balance on track.
- Careful attention must be paid to wing supports to ensure that they are adequate to resist the maximum forces and control wing dynamic oscillations.

EXAMPLE 9.7

Using the values in *Table 9.3* estimate the downforce and drag on a single-element wing of span 1.3 m and chord length 0.4 m on a car travelling at 50 m/s. How much engine power does the wing consume?

Equation [9.12]

$$\text{Downforce} = \tfrac{1}{2} C_L \rho v^2 A = \tfrac{1}{2} \times 1.2 \times 1.204 \times 50^2 \times 1.3 \times 0.4$$

$$= \mathbf{939\ N}$$

Equation [9.13]

$$\text{Drag force}, F_D = \tfrac{1}{2} C_D \rho v^2 A = \tfrac{1}{2} \times 0.3 \times 1.204 \times 50^2 \times 1.3 \times 0.4$$

$$= \mathbf{235\ N}$$

$$\text{Power consumed} = F_D \times v = 235 \times 50/10^3$$

$$= \mathbf{11.8\ kW}\ (15.5\ \text{bhp})$$

9.3.4 Front and rear wing issues

Front wings generally operate in clean air with laminar flow (unless closely following another car) and this is conducive to good downforce performance and wing efficiency. It is usually easier to get front downforce than rear. An important issue is the height above the ground at which the wing is mounted. When the gap between the wing and the ground is less than say 50% of the chord length, air is accelerated as it is funnelled down the gap between the underside of the wing and the ground. This increase in speed of the air produces a further reduction in pressure on the underside of the wing in compliance with the Bernoulli equation. The wing is said to be operating in **ground effect**. The resulting additional downforce may be beneficial but it can also cause problems. The effect can be particularly marked during high-speed braking when load transfer to the front of the car further reduces the front wing ground clearance. It is possible that the front wing is sucked down to the extent that the end-plates touch the ground and /or flow is blocked by the boundary layer at which point downforce is reduced. Under these conditions pitching oscillations or 'porpoising' can occur and the driver needs to cope with rapidly varying downforce at the front of the car. The usual remedy for this is to mount the front wing higher to reduce the ground effect.

Rear wings generally operate in dirtier air after it has been disturbed passing over the bodywork of the car. Consequently it is beneficial to mount the main rear wing as high as regulations permit. The lift coefficients given in *Table 9.3* assume a wing acting in isolation in clean air and consequently it is prudent to be generous with the rear wing size at the initial design stage to allow for possible inefficiencies. In addition to the main wing it is common to provide a small lower wing just above the exits from the rear diffusers and this will be covered in the next section.

9.4 The floor-pan

Because wings depend upon a significant angle of attack to deflect the air flow they inevitably incur a substantial drag penalty. Downforce from the floor-pan however is primarily based upon creating a large low-pressure area under the car by, as the Bernoulli equation indicates, increasing the velocity of flow between the floor-pan and the road. There is much less drag associated with this approach and hence it makes sense to maximise the downforce derived from the underbody.

The initial approach to underbody aerodynamics was to consider the base of the side-pods as curved inverted wings with side-skirts to contain the low pressure; however as regulations became more restrictive an alternative view emerged. Many formulae prescribed the use of flat or stepped floors within the wheelbase of the car and a minimum ground clearance (usually about 40 mm) is a requirement. These factors pointed towards viewing the situation as an example of the well-known **Venturi tube**, the workings of which are explained by the Bernoulli equation. A **Venturi** is a device often used for measuring flow in a pipe and consists of a section which funnels fluid into a narrow orifice and then expands it again to the original pipe diameter. As the flow speeds-up through the orifice the static pressure drops. *Figure 9.9* indicates a typical implementation on a racing car and shows a rough indication of the pressure distribution under the car.

The purpose of the **infuser** is to funnel airflow under the car, however some formula regulations do not allow such variations in floor profile and designers may be forced to keep the floor horizontal at the front. With a well-designed floor-pan and diffuser this may not be too much of a problem as air will curve-down to flow under the car. It can be seen from *Figure 9.9* that there is a small area of positive pressure at the start of the infuser which clearly reduces the effective downforce a little. The slope of the infuser must be low enough to maintain attached laminar flow. The flow is in a favourable pressure gradient – high to low – so 15° is probably typical.

The **throat** is the Venturi orifice and the main area over which the negative pressure acts and hence its area should be maximised. Note that in *Figure 9.9* the floor is shown sloping upwards towards the rear of the car, known as **rake**. At first sight this may seem wrong as the purpose of the throat is to maximise the speed of the flow; however the clue to why this is done is in *Example 9.4* which is concerned with the thickness of the boundary layer. The example suggests that at the rear of the car the boundary layer has become turbulent and is of the order of 46 mm thick. This means that if the floor was at the constant ground-clearance height of say 40 mm throughout, by the time the flow got to the rear of the car, it would be choked by the slow moving boundary layer and the static pressure could become positive. A rake of 1°–1.5° is generally adopted. It can be seen that there are downforce peaks at the junctions between the throat and the infuser and the diffuser. This is explained by change in momentum as the flow is pulled around the corner. Where possible the edge of the floor should terminate in a vertical lip or skirt which should extend down to the minimum permitted ground clearance to contain the flow and reduce negative pressure leakage to a minimum.

The **diffuser** performs three main functions. Firstly, it produces some downforce in its own right provided its slope angle is small enough to maintain attached flow. Secondly, it slows down the flow to that of the surrounding air before it leaves at the back of the car. Thirdly, and most importantly, it exits the car at the rear which is a negative pressure zone and this further speeds-up the flow under the car. The diffuser hence acts like a vacuum pump drawing air in at the front of the car. Flow is against an adverse pressure gradient – low

Figure 9.9
Floor-pan aerodynamics

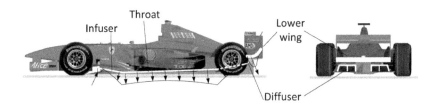

to higher – and to avoid separation a relatively gradual slope of about 7° is desirable, however this can be increased if a lower rear wing is used.

The **lower rear wing** is positioned just above and behind the diffuser exit. We saw in *Figure 9.6a* that a wing produces a significant negative pressure zone on its underside and this serves to improve the pressure gradient of the emerging diffuser flow. It gives the air 'something to aim at'. Under such circumstances diffuser angles can be increased significantly to say 20°. Angles of 30° are used in F1 although additional measures are taken, such as using the exhaust gases to energise the flow with the so-called 'blown diffuser'.

Later consideration of the aerodynamic balance of a car requires knowledge of where the resultant force from the underbody pressures occurs. This requires consideration of the plan shape of the floor and the pressure distribution which is likely to change somewhat as the velocity increases.

9.5 Other devices

Whereas the wings and floor-pan produce the bulk of the downforce, a modern racing car contains many additional aerodynamic devices such as winglets and turning vanes whose job it is to manage and optimise the flow of air as it passes over the car. The following is a brief description of some of them – all of which have aspects in common:

Barge boards

Barge boards are curved plates normally mounted on each side of the car a few centimetres from the bodywork sides in front of the sidepods – see *Figure 9.1*. They are curved to deflect air flow away from the centre-line of the car. Their main purpose is to separate the smooth flow along the sides of the car from the turbulent wake behind the front wheels. They thus help to maintain laminar flow into the sidepod radiator openings. The bottom edge of a barge board has become important as a vortex generator. Barge boards can induce significant changes even at the rear of the car however (*ref. 14*).

Vortex generators

The creation of turbulent flow in the form of spiral vortices can be beneficial. A **vortex generator** usually takes the form of a plate mounted parallel to the direction of flow. The pointed tip of the device induces a trailing 'tornado like' twisting vortex. The high-speed rotating flow around the vortex core is at low pressure. Such devices are often used within infusers to create vortices that snake under the car and have been found to increase underbody downforce. They also have a role when mounted in such a way as to produce vortices along the side edges of flat floors. These vortices act as skirts to separate the low-pressure zone under the floor from the surrounding air. With aircraft the main application of vortex generators is as small devices on the low-pressure side of wings. They poke just beyond the boundary layer and take energy from the passing flow to re-energise the boundary layer and delay separation. This application is now being used on cars (*ref. 11*).

Strakes

Strakes can take two forms. They are firstly curved vertical plates often mounted in diffusers or under front wings to guide the airflow. Secondly, they can be small sloping winglets designed to produce a small additional downforce contribution. When mounted at the sides of the nose they are referred to as 'dive plates'. Although inclined flat plates will produce downforce they are never as efficient as aerofoil sections as the flow becomes separated on the low-pressure side.

9.6 Aerodynamic balance

Figure 9.10 shows the principal aerodynamic resultant forces together with estimates of their position. Drag forces associated with the front wing and the underfloor exist but are both small and close to the ground compared to the rear wing and so can be ignored. The objective is to strive for a car where the understeer/oversteer balance is appropriate and constant at all speeds. This may not be fully achievable as the relative magnitudes and positions of the aerodynamic forces may change with speed. Slow speed balance is dominated by mechanical balance as discussed in *Chapter 5*. It follows that the effect of aerodynamic forces on the front/rear wheel load distribution should maintain the original mechanical static weight distribution. Thus if a car has a 45:55 front/rear weight distribution, then the initial aim should be for the

Figure 9.10
Aerodynamic balance

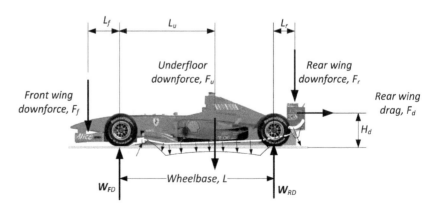

wheel load increases from aerodynamic downforce to be in the same ratio. Final adjustment can then be made on the circuit by adjusting the angles of attack.

From *Figure 9.10*

Take moments about front axle for rear axle downforce, W_{RD}:

$$W_{RD} \times L = F_u \times L_u + F_r \times (L + L_r) + F_{dr} \times H_d - F_f \times L_f$$

$$W_{RD} = [F_u L_u + F_r(L + L_r) + F_{dr} H_d - F_f L_f]/L \qquad [9.14]$$

Take moments about rear axle for front axle downforce, W_{FD}:

$$W_{FD} \times L = F_f \times (L + L_f) + F_u \times (L - L_u) - F_r \times L_r - F_{dr} \times H_d$$

$$W_{FD} = [F_f(L + L_f) + F_u(L - L_u) - F_r L_r - F_{dr} H_d]/L \qquad [9.15]$$

EXAMPLE 9.8

Given the following data in the context of *Figure 9.10* determine the aerodynamic downforce ratio between the front and rear axles.

F_f	F_u	F_r	F_{dr}	L	L_f	L_u	L_r	H_d
0.950 kN	0.975 kN	1.140 kN	0.360 kN	2.300 m	0.550 m	1.600 m	0.400 m	0.750 m

Equation [9.14]

$$W_{RD} = [F_u L_u + F_r(L + L_r) + F_{dr} H_d - F_f L_f]/L$$

$$= [(0.975 \times 1.6) + (1.140 \times (2.3 + 0.4)) + (0.36 \times 0.75) - (0.95 \times 0.55)]/2.3$$

$$= [1.56 + 3.08 + 0.27 - 0.52]/2.3$$

$$= 1.91 \text{ kN}$$

Equation [9.15]

$$W_{FD} = [F_f(L + L_f) + F_u(L - L_u) - F_r L_r - F_{dr} H_d]/L$$

$$= [(0.95 \times (2.3 + 0.55)) + (0.975 \times (2.3 - 1.6)) - (1.140 \times 0.4) - (0.36 \times 0.75)]/2.3$$

$$= [2.71 + 0.68 - 0.46 - 0.27]/2.3$$

$$= 1.16 \text{ kN}$$

Check vertical equilibrium:

$$\text{Total downforce} = 0.950 + 0.975 + 1.140 = 3.065 \text{ kN}$$

$$\text{Total vertical axle loads} = 1.91 + 1.16 = 3.07 \text{ kN} \checkmark$$

$$\text{Front/rear ratio} = 38{:}62$$

Race car design

> **Comment:**
> *If the ratio is considered to be too biased towards the rear, the angle of attack of the rear wing could be reduced assuming the front wing is already at its maximum angle of attack.*

9.7 Design approach

The following steps are suggested for the initial design of the aerodynamic package:

1. Maximise the contribution from the floor-pan as this carries little drag penalty.
2. Estimate the amount of engine power that can be devoted to overcoming the aerodynamic drag from the wings. To do this firstly determine the required maximum speed of the car, v_{max}, in m/s. This should be say 10% more than the expected top speed in a race to allow for acceleration. Next estimate the required power at the wheels in order to achieve this for a car without wings. *Table 9.2* suggests a drag coefficient of 0.6 for an open-wheeled car without wings:

Equation [9.9]
$$\text{Drag force (no wings)}, F_D = \tfrac{1}{2} C_D \rho v^2 A$$
$$= \tfrac{1}{2} \times 0.6 \times 1.204 \times v_{max}^2 \times A$$
$$= 0.36 A v_{max}^2 \text{ N}$$

where A is the frontal area of the car.

$$\text{Power consumed (no wings)} = F_D \times v = 0.36 A v_{max}^3 \times 10^{-3} \text{ kW} \quad [9.16]$$

The above approach ignores the power consumed in rolling resistance of the tyres. This value is then deducted from the total power available at the wheels to obtain the power, P_{drag}, that is available for overcoming aerodynamic drag from the wings.

3. Divide the power figure, P_{drag}, by the maximum speed to give the maximum total wing drag, F_d. Assuming that the rear wing contributes say 50% of wing drag:

Equation [9.13]
$$\text{Rear wing drag force}, F_{dr} = 0.5 F_d = \tfrac{1}{2} C_D \rho v_{max}^2 A$$

where A is the plan area of the wing. The unknowns in the above equation are the drag coefficient C_D and the area A. If a trial value for A

(span × chord) is estimated the maximum value of drag coefficient can be evaluated. *Table 9.3* can then be consulted and the number of elements selected. Its downforce can then be calculated:

Equation [9.12]
$$\text{Rear wing downforce, } F_r = \tfrac{1}{2} C_L \rho v_{max}^2 A$$

where A is again the plan area of the wing.

4. Make assumptions about the magnitude and location of the underfloor downforce and then perform the aerodynamic balance calculation as in *Example 9.8* to determine the required front wing downforce. Again use *Table 9.3* to estimate the type and size of front wing. Check that its drag is within the 50% allowed for the front wing.

EXAMPLE 9.9

The following data (as defined in *Figure 9.10*) is typical of a FSAE/Formula Student car. Establish initial proposals for front and rear wings and estimate the total downforce at 20 m/s which is a typical speed in the endurance event.

F_u (20 m/s)	L	L_f	L_u	L_r	H_d
0.2 kN	1.6 m	0.5 m	0.9 m	0.4 m	0.8 m

Max. speed	Frontal area (no wings)	Power at wheels	Front/rear weight distribution
45 m/s*	0.80 m²	60 kW	45:55

* Relatively high top speed required for the acceleration event.

Equation [9.16]

$$\text{Power consumed (no wings)} = 0.36 A v_{max}^3 \times 10^{-3} \text{ kW}$$
$$= 0.36 \times 0.8 \times 45^3 \times 10^{-3} \text{ kW}$$
$$= 26.2 \text{ kW}$$
$$P_{drag} = 60 - 26.2 = 33.8 \text{ kW}$$

Total wing drag force, $F_d = 33.8/45 = 0.75$ kN

Rear wing drag force, $F_{dr} = 0.5 F_d = 0.5 \times 0.75$

Race car design

Equation [9.13] $= 0.37 \text{ kN}$ $= \tfrac{1}{2} C_D \rho v_{max}^2 A$

Try a wing of say 1.2 m span × 0.5 m chord:

$$C_D = 2 \times 0.37 \times 10^3 / \rho v_{max}^2 A$$
$$= 0.72 \times 10^3 / (1.204 \times 45^2 \times 1.2 \times 0.5)$$
$$= \mathbf{0.50}$$

Inspection of *Table 9.3* indicates that the above C_D value lies between that for single- and two-element wings. It is suggested that a two-element wing operating at an angle of attack somewhat below its maximum is selected. A two-element wing at say 16° angle of attack would be expected to have a C_D of about 0.5 and a C_L of about 1.6. (An alternative would be a bigger single-element wing.)

Downforce and drag at 20 m/s

Equation [9.12]

Rear wing downforce, $F_r = \tfrac{1}{2} C_L \rho v_{max}^2 A = \tfrac{1}{2} \times 1.6 \times 1.204 \times 20^2 \times 1.2 \times 0.5$
$$= 231 \text{ N} \qquad = \mathbf{0.231 \text{ kN}}$$

Equation [9.13]

Rear wing drag force, $F_{dr} = \tfrac{1}{2} C_D \rho v_{max}^2 A = \tfrac{1}{2} \times 0.5 \times 1.204 \times 20^2 \times 1.2 \times 0.5$
$$= 72 \text{ N} \qquad = \mathbf{0.072 \text{ kN}}$$

Determine the required front downforce for 45:55 weight distribution:

Equation [9.14]

$$W_{RD} = [F_u L_u + F_r(L + L_r) + F_{dr} H_d - F_f L_f]/L$$
$$= [(0.2 \times 0.9) + (0.231 \times (1.6 + 0.4)) + (0.072 \times 0.8) - (F_f \times 0.5)]/1.6$$
$$= [0.18 + 0.46 + 0.06 - 0.5 F_f]/1.6$$
$$= 0.44 - 0.31 F_f \text{ kN} \qquad ①$$

Equation [9.15]

$$W_{FD} = [F_f(L + L_f) + F_u(L - L_u) - F_r L_r - F_{dr} H_d]/L$$
$$= [(F_f \times (1.6 + 0.5)) + (0.2 \times (1.6 - 0.9)) - (0.231 \times 0.4) - (0.072 \times 0.8)]/1.6$$
$$= [2.1 F_f + 0.14 - 0.09 - 0.06]/1.6$$
$$= -0.01 + 1.31 F_f \text{ kN} \qquad ②$$

But $W_{RD} = W_{FD} \times 55/45 = 1.22 W_{FD}$ ③

and substituting ① and ② into ③:

$$0.44 - 0.31F_f = 1.22(-0.01 + 1.31F_f) = -0.01 + 1.60 F_f$$
$$0.45 = 1.91F_f$$
$$F_f = 0.236 \text{ kN}$$

Specify front wing:

It can be seen that the downforce required for the front wing is almost identical to that obtained for the rear wing (0.231 kN); therefore use the same wing type and dimensions.

Use two-element wings with 1.2 m span and 0.5 m chord at 16° angle of attack front and rear.

As the wings are the same the 50% drag allowance at 45 m/s will be satisfactory.

Comments:
1. *If regulations permit, the angles of attack of front and rear wings could be increased to say 20° for events where the maximum speed of 45 m/s is not achieved.*
2. *The total predicted downforce at 20 m/s, including the floor, is about 650 N which is adding about 24% to the weight of the car plus driver. A rough tyre analysis suggests that this may add about 16% to the cornering grip, raising the maximum lateral g from say 1.5 to 1.75. This is definitely worth having. The downside is extra weight and drag which will adversely affect acceleration a little.*

SUMMARY OF KEY POINTS FROM CHAPTER 9

1. **Bernoulli's equation** indicates that when the speed of airflow is increased the pressure drops and this important principle is widely exploited to create downforce.
2. Viscous fluid flows, such as air, attach themselves to solid surfaces creating a **boundary layer** of slow-moving particles. This causes **skin-friction drag**.
3. **Form drag** results from the shape and frontal area of objects. It is a force which acts opposite to the direction of motion and is the result of surface pressure. Total drag increases with the square of the flow velocity and is quantified by means of a **drag coefficient, C_D**.
4. Air has mass and changing its **momentum** by altering its direction requires a force. This is the origin of downforce and drag from wings and other sloping surfaces.
5. **Wings** generate downforce efficiently because the upper surface has positive pressure and the lower surface negative pressure. Downforce and drag increase as the **angle of attack** increases, but at a certain point the flow separates from the lower surface and some downforce is lost.
6. Multiple-element wings can operate at higher angles of attack and generate more downforce and drag.
7. Air passing under the floor of a car can provide a similar amount of downforce to a wing. This can be compared to a Venturi tube which accelerates air through an **infuser** into a narrow **throat** and finally expands it again in a diffuser. Bernoulli indicates that the pressure drops as the air speeds up in the throat.
8. Designers should aim for a **balanced** aero package that maintains good understeer/oversteer handling as the speed of the car increases.

10 Engine systems

> **LEARNING OUTCOMES**
>
> At the end of this chapter:
> - You will understand the principles behind the four-stroke petrol engine
> - You will be familiar with the systems that are required to support the effective working of the engine
> - You will understand the importance of tuning the air induction and exhaust systems in order to maximise the engine's performance
> - You will be aware of the issues of introducing a restrictor into the air supply
> - You will be aware of the problems of fuel and oil surge and how they can be ameliorated

10.1 Introduction

This book does not cover the design of the internal components of engines; however it does provide a brief introduction to several key external systems to enable the engine to function. For a four-stroke petrol engine these include:

- air induction,
- fuel induction,
- exhaust,
- engine management and ignition,
- cooling,
- lubrication.

Each of these will be considered briefly, largely in the context of a motorcycle-engined racing car. The aim is to highlight the major design issues and to provide pointers to further work.

Firstly, for those new to engines, the basics of the **four-stroke cycle** will be considered. A car four-stroke petrol engine typically has between one and twelve **cylinders** arranged either in a straight line or in a 'V' formation. Each cylinder contains a sliding **piston** which is connected to a rotating **crankshaft** by a **connecting rod**. This arrangement converts the sliding motion of the piston into rotary motion of the shaft. *Figure 10.1* illustrates the four cycles:

Figure 10.1
The four-stroke cycle

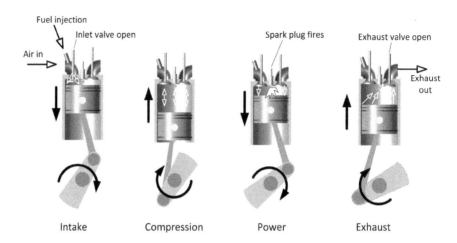

- **Intake cycle** – As the piston descends, the **inlet valve** opens and air is drawn into the cylinder. At the same time the fuel injector opens and a jet of petrol is squirted in. When the piston reaches the bottom of the cylinder the inlet valve is closed to form a sealed volume.
- **Compression cycle** – The rising piston compresses the air and petrol mixture.
- **Power cycle** – The air and petrol mixture is ignited by the **spark plug**. The resulting explosion forces the piston down, generating power.
- **Exhaust cycle** – As the piston rises again the **exhaust valve** opens and the combustion products are forced out of the **exhaust valve**.

It follows from the above that the crankshaft must revolve twice for each set of four cycles. Multiple cylinder engines are smoother if the power cycles are staggered according to a **firing order**. Many straight-four-cylinder motorcycle engines adopt a 1–2–4–3 firing order, counting from one end of the engine. The inlet and exhaust valves are commonly opened and closed by rotating **camshafts**.

10.2 Air induction – the normally aspirated engine

The normally aspirated engine relies upon a partial vacuum created by the descending piston to draw air into the cylinder. However a modern high-revving engine is a highly dynamic system with many subtleties to improve efficiency and power output. For example, at the end of the exhaust cycle, the exhaust valve does not close until *after* the inlet valve has opened to start the inlet cycle. This is known as **valve overlap** and exploits the fact that the out-rushing exhaust gases suck in more incoming air than would otherwise be the case. The cylinder has become **overcharged**. This means that more petrol can be added and more power generated. The degree of over-charging is expressed as the **volumetric efficiency** and a key aim in **tuning** an engine is to maximise the volumetric efficiency. Careful design of the air induction and exhaust systems also contribute to volumetric efficiency and values of over 120% are achievable.

Figure 10.2 indicates the principal elements of an air intake system. The **inlet trumpet** is designed to provide smooth entry to the system and if oriented to point towards the front of the car can increase volumetric efficiency through **ram-air**, although the effect is small with vehicle speeds below 50 m/s. The basis of ram-air is the conversion of dynamic pressure into static pressure in accordance with the Bernoulli equation. The outside diameter of the trumpet must be greater than that required to provide sufficient air flow to the engine at the target speed. As the airflow strikes the inlet it is slowed down with an accompanying rise in pressure. Ideally the tube following the inlet, but before the restrictor, should be a slowly expanding volume to further slow the flow and increase the pressure even further.

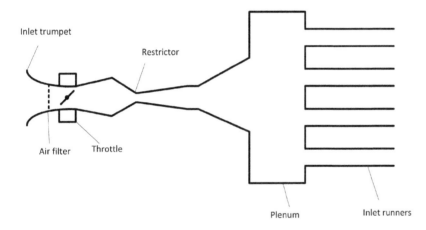

Figure 10.2
The air induction system

The **throttle** is the means by which the driver, via the accelerator pedal, controls the air flow and hence the speed of the vehicle. The throttle body contains an angle sensor that is used by the engine management unit to determine the fuel requirements.

The **restrictor** is a requirement of many formulae and they are usually defined in terms of the maximum diameter through which all air must pass. They provide a very firm limit on the peak power output of the engine because the velocity of air through the restriction is limited to the speed of sound (343 m/s at 20°C). As engine revs increase the increased demand for air to fill the cylinders causes a greater pressure difference between the inlet and outlet sides of the restrictor. This causes the flow velocity through the restrictor to increase, but only up to the point where the velocity of sound is reached. Any further reduction in downstream pressure will not increase the flow rate and the flow is said to be **critical** or **choked**. (Although the flow velocity is limited, the mass flow rate can still be increased by compressing the inlet air so that it is denser – see **forced induction** in *section 10.3*.) At low engine speeds the restrictor will have no effect on power output, but as the speed of sound is approached the peak power will level off. The shape of the restrictor requires careful attention to maximise the flow characteristics as the choking starts to take effect. A Venturi tube has been found to be an efficient shape to maximise mass flow through a restriction. *Plate 7* shows an example of a Venturi that has been investigated by a computer-based **computational fluid dynamics (CFD)** package to optimise the shape, where the objective is to maximise the mass flow rate at high flow velocities. This is equivalent to minimising the pressure drop across the restrictor.

The **plenum**, **intake chamber** or **air box** provides a resonating volume of air downstream of any restrictor. In a **tuned** system this works in conjunction with the **inlet runners** to increase the volumetric efficiency of the engine. The basis of such tuning is to exploit the elastic nature of air to create shock waves that 'bounce' an additional charge of air into the cylinder just before the inlet valve closes. Resonating waves can take the form of either **compression** or **expansion** waves. When a wave encounters a hard stop, such as a closed valve, it is reflected in the same form. However if a wave encounters an open end, such as in the plenum, it is reflected in the opposite form – i.e. a compression wave is reflected as an expansion wave. A typical process is as follows:

1. An inlet valve opens and the piston descends causing a low-pressure expansion wave to propagate up the inlet runner to the plenum.
2. When the wave reaches the plenum opening an opposite high-pressure compression wave is reflected back down the inlet runner to the valve just before it is about to close.
3. An extra charge of high-pressure air enters the cylinder and is trapped by the closing valve.

It is clear from the above that the wave pulses must be coordinated with valve action. The time taken by the wave depends upon the length of the inlet runners and the valve timing depends upon the speed of the engine. It follows that the length of the inlet runners is optimised to increase the volumetric efficiency at a particular engine rpm. The downside is that the pulsed waves can produce a *reduction* in volumetric efficiency at other engine rpm. The designer therefore must decide which engine speed to target for the boost in torque and power. The exhaust system is tuned in a similar way to generate a boost in volumetric efficiency from pulsed waves. The two extreme choices are:

1. Optimise the length of inlet and exhaust pipes to concentrate maximum volumetric efficiency at the top-end of the rpm range. This will maximise the headline power output of the engine but produce a 'peaky' engine

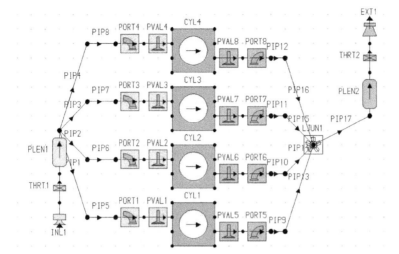

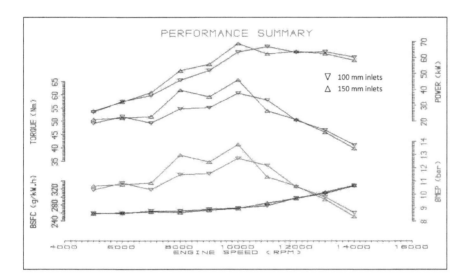

Figure 10.3
Lotus Engine Simulation model and results (Lotus Engineering, Norfolk, England). Unrestricted 600 cc four-cylinder motorcycle engine

with reduced bottom-end torque. The driver will need to rev the engine hard and make frequent gear changes to keep within the power band.

2. Optimise the length of inlet and exhaust pipes to distribute the individual peaks from the inlet and exhaust pipes throughout the mid-range engine rpm, taking care to avoid a mid-range 'trough'. This produces a more driveable car with less gear changing and wear-and-tear on the engine but lower peak power.

The experienced driver might prefer a solution closer to the first option but the beginner closer to the second.

Hand techniques, such as the Helmholtz resonator method (*ref. 26*), exist to estimate the inlet runner lengths; however they invariably involve simplifications such as assuming the speed of the shockwaves is constant. More accurate results, particularly for multi-cylinder engines, are obtained using modern software packages such as Lotus Engine Simulation (*ref. 13*). *Figure 10.3* shows an example of a Lotus Engine Simulation graphical model for a four-cylinder motorcycle engine with both the inlet and exhaust systems shown. Also shown are the simulation results for two alternative lengths of inlet runner – 100 mm, shown as inverted triangles, and 150 mm, shown as triangles. The observable peaks in the torque curves are the result of pulsed wave tuning. Note that the 100 mm inlet runners produce peak power at 11 000 rpm whereas the 150 mm runners peak at 10 000 rpm. The graphs also indicate *brake specific fuel consumption* (BSFC), which is concerned with fuel efficiency, and *brake mean effective pressure* (BMEP), which is related to torque.

10.3 The forced induction engine

The previous section considered air entering the engine under atmospheric pressure plus the addition of pulsed shockwaves. The torque, and hence power, output of an engine can be increased considerably if the air is compressed and enters the engine under pressure – known as **forced induction** (of course, the extra air must be accompanied by extra fuel). This is achieved by the use of either a **supercharger** or a **turbocharger**:

- A **supercharger** is usually driven by a belt and pulley from the engine. The degree of 'boost' or pressure increase is directly related to engine speed. Superchargers are relatively simple to install and manage but can only be adjusted by changing the pulley diameters.
- A **turbocharger** is driven by engine exhaust gases. Turbochargers can be designed to produce peak boost at relatively low engine speed which is then sustained up to the rev limit. Hence turbochargers can produce more average power over the rev range. Excess pressure is controlled by a pressure relief valve known as a **wastegate**. The use of exhaust gases can cause heat problems and may require the use of an **intercooler**. These are more complex to install and manage and often introduce a delay between the driver pressing the accelerator and the actual power boost – known as **turbo-lag**.

Many formulae either forbid the use of forced induction or require cars with it to compete against naturally aspirated cars with bigger engines. Formula SAE/Student regulations currently permit forced induction but the device must be placed *after* the restrictor. This means that the mass flow rate, and hence peak power, cannot be increased compared to the naturally aspirated engine, although performance at lower engine speeds can be improved. Incidentally, forced induction *before* the restrictor *can increase* peak power as, despite the speed-of-sound limit still applying, air density and hence mass flow rate increase.

Chapter 10 Engine systems

10.4 Fuel Induction

The basic elements of the fuel system are shown in *Figure 10.4*.

The **fuel tank** can either be a rigid, often aluminium alloy, container or a flexible polymer bag or **bladder**. Bladders are often placed inside a rigid container for protection. There generally needs to be some form of system to prevent **fuel surge**. This occurs at low fuel levels when the lateral cornering force moves the fuel to one side of the tank away from the **collector** that supplies the pump. Under such circumstances the pump can supply air to the fuel injectors. The driver will experience engine splutter and lack of power on long corners. The two main solutions are either the use of internal baffles with one-way valves to prevent lateral movement (see *Figure 10.4*), or the provision of a separate **swirl pot** between the tank and the pump. This is a small tank that is kept filled by an additional low-pressure pump which provides a reservoir for the main high-pressure pump. The fuel tank can be filled with special foam. Although this helps combat fuel surge a little, its main purpose is to prevent explosion and rapid discharge of fuel in the event of a crash.

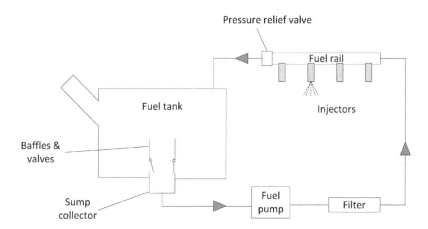

Figure 10.4
Fuel system

The main **fuel pump** must be of the high-pressure type for fuel injection engines. The pump may be contained within the fuel tank as on most road vehicles. The **fuel rail** is generally held at a pressure of about 3 bar by the **pressure regulator** and surplus fuel is returned to the tank. The fuel injectors spray fuel into the cylinders and are simply turned-on for the appropriate duration by the engine control unit (ECU). Injectors are available with different flow rates and spray patterns. They should be installed in the inlet runner pipes so that the fuel sprays directly into the cylinder.

All fuel system installations should be 'plumbed' with fuel-approved pipework and fittings. This is particularly important on the high-pressure side of the pump.

10.5 Exhaust system

As already indicated, the exhaust system is an important element for tuning the engine. The pipes induce pulsed shock waves that affect the volumetric efficiency of the engine in a similar way to the inlet runners. The pipes closest to the engine are known as **primaries** and a practical challenge is that they should all be of the same (and correct) length to provide uniform tuning between cylinders. *Figure 10.5* shows all four primaries connected to a single **secondary** via a manifold. This is known as a 4 into 1 system. An alternative for four-cylinder engines is 4 into 2 into 1. In this case the length of the two secondaries can also be tuned to provide yet another volumetric efficiency peak. A typical motorcycle engine will fire in the order 1–2–4–3 and, in this case, the primaries from cylinders 1 and 4 and cylinders 2 and 3 would be joined to form secondaries. This avoids wave interference from cylinders that fire immediately after each other.

It is desirable to wrap the exhaust system with heat-resistant lagging. This improves engine efficiency and helps to keep the engine bay cooler.

In most formulae the exhaust system must culminate with a **silencer** that is crucial to meeting the maximum sound emissions requirements. Silencer design can significantly influence engine power output and the only convincing way to prove this is through the production of torque/power curves on an engine dynamometer.

Figure 10.5
Typical 4 into 1 exhaust system

10.6 Engine management and ignition

The primary functions of the **engine management system** are to fire the spark plugs at the appropriate time (ignition timing) and to control the amount of fuel injected into the engine by determining the period for which

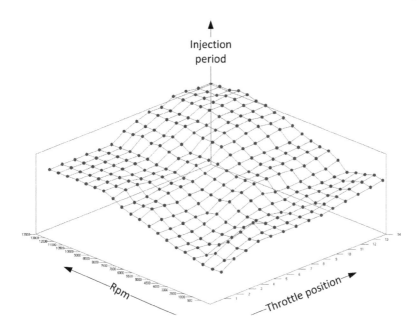

Figure 10.6
Engine fuel map (produced with software from DTAfast)

each injector is activated. It does this by reading various engine sensors such as a crank sensor for engine rpm, a cam sensor to determine the engine cycle and a throttle position sensor for the volume of air entering the engine. Armed with this information, the system uses look-up tables of programmed data to determine the outputs to the plugs and injectors. The look-up table data is often presented in the form of 3-D graphs and consequently they are often referred to as **engine maps**. *Figure 10.6* shows a typical **fuel map**. The horizontal axes represent rpm and throttle position and the vertical axis represents the time period during which the fuel injectors are open. A water-temperature sensor modifies the map to permit cold starting. Modern road cars utilise many more sensors to optimise economy and emissions.

It is possible to use a standard engine ECU to control a race engine, however this often takes the form of a sealed 'black box' which is difficult to reprogram to suit modified induction and exhaust systems. The standard maps will not be appropriate, particularly if a restrictor is used. One option is to combine the standard ECU with a device such as a Power Commander (*ref. 20*). This is a commercially available programmable device that sits between the standard ECU and the fuel injectors to modify the fuel map. This requires the engine to be tuned on a dynamometer or rolling road.

A better option to the standard engine ECU is the adoption of a stand-alone specialist performance **engine management system**. Many are available to suit a wide range of budgets from companies such as DTA (*ref. 8*), Emerald (*ref. 9*) and Motec (*ref. 16*). In addition to easy programming of the ignition and fuel maps, these systems provide many additional features that are useful in racing such as **launch control**, **traction control** and **shift-cut** for clutch-less gear-changes.

The general aim of engine tuning is to program the ECU so that it maintains an optimum target air/fuel mixture under most running conditions. Combustion involves a reaction between oxygen molecules and fuel molecules. Knowing that air contains about 23% of oxygen we can define a ratio (by weight) of air/fuel such that, after combustion, there will be no molecules of either oxygen or air remaining (in theory). This is known as the **stoichiometric ratio** and is 14.68 parts (by weight) of air to one part of premium petrol fuel (gasoline). Another way of expressing this ratio is the lambda (λ) excess oxygen ratio. A value of $\lambda = 1.0$ represents the stoichiometric ratio.

- Most car manufacturers aim for a value of $\lambda = 1.0$ as this provides a good compromise between economy and emissions.
- A value of $\lambda > 1.0$ means excess oxygen and a **lean** mix. A lean mix with a ratio of 15.4 ($\lambda = 1.05$) is best for fuel economy but also produces more nitrous oxide emissions and increased heat in the exhaust system.
- A value of $\lambda < 1.0$ means excess fuel and a **rich** mix. A rich mix with a ratio of 12.6 ($\lambda = 0.95$) is best for power output and hence this is the target for racing.

The above points are summarised in *Figure 10.7*.

A vital aid for programming ECU devices is the wide-band lambda (λ) oxygen sensor. These sensors are placed in the exhaust system close to the engine. Ideally there should be one in each exhaust header but generally only one is positioned at the point where the exhaust tubes join into one. The lambda sensor is interfaced with the ECU and can be used to adjust the air/fuel mixture in real time as the car is running – known as **closed loop programming**. However generally, for racing, the ECU is programmed in **open loop** with the engine on a dynamometer or, better still, the complete car on a rolling road. The car must be complete with the final induction and exhaust system.

Figure 10.7
The air/fuel mixture

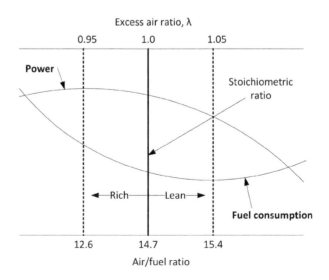

A skilled tuner will exercise the engine over its torque and rpm range and adjust the fuel injection periods to maintain the target lambda value.

10.7 Cooling

Internal combustion engines are only about 30% efficient in converting the energy content of fuel into useful mechanical work. About 15% of the fuel energy is lost by unconverted chemical energy and kinetic energy passing through the exhaust. The remaining 55% is converted into heat which has to be dissipated through cooling in the water and oil systems, the exhaust and by radiation from the engine surfaces. *Table 10.1* shows roughly how this 55% is distributed between the various heat dissipation systems. These approximate percentages will vary somewhat depending upon how hard the engine is working. The relevant engine power output value, in this case, is the *average* power output at the flywheel over a lap. This can be estimated at about 50% of the peak engine power, but as we saw in *Example 9.5*, the relatively high drag coefficients of racing cars require significant power just to maintain high speeds, so this value could rise for fast circuits. Hence for a car with a relatively modest 160 kW of peak engine power it is suggested that the average power output over a lap is say 80 kW and that this uses 30% of the fuel energy. Hence the 19% (*Table 10.1*) dissipated by the water-cooling system represents 80 × 19/30 = 50 kW of heat! This is clearly a significant amount when compared to say a domestic room heater which has about 3 kW output. For Formula SAE/Student cars the average power output over an endurance lap is probably in the 10–20 kW range, resulting in say 6–12 kW to be dissipated by the water-cooling system.

Table 10.1 Heat dissipation through different systems

System	% of heat dissipated
Water	19%
Oil	6%
Exhaust	25%
Radiation	5%

Figure 10.8 shows a typical water-cooling circuit. The system is filled via the **header tank** which is located at the highest point in the system. This is fitted with a spring-loaded pressure relief cap and as the heated water expands any excess is expelled to the **catch can**. The **thermostat** facilitates quicker engine warm-up as it only allows water to circulate through the circuit once it has reached a specified temperature.

Figure 10.8
Water-cooled circuit

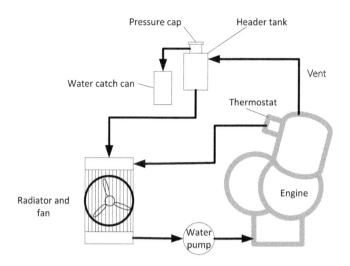

The **radiator** is the main heat exchange component. It is possible to perform detailed heat transfer calculations to size the radiator, hoses and fan; however the best starting point is to base the sizes on an existing system from an engine with a similar power output (not cubic capacity). The aim should be to achieve a maximum water temperature in a race of about 90°C. Ideally the radiator should be enclosed in a close-fitting air duct. The entry to the duct, which is often in a side-pod, should face the front of the car and be roughly 30–50% of the radiator frontal area. After entry, the duct should expand slowly to the size of the radiator. This slows the air and, from Bernoulli, increases the pressure before passing through the radiator. After the radiator the duct area should reduce to return the air to the speed of the surrounding air and exit in a low-pressure zone at the rear of the car. This arrangement maximises the effectiveness of the radiator and minimises aerodynamic drag.

It is possible to retain the original mechanical **water pump**; however there are some advantages in replacing it with an electric pump. Electric pumps take less power from the engine at high rpm and can be left running at the end of a race when the engine is very hot. Also, with many motorcycle engines, removing the mechanical water pump provides an ideal location for a dry sump scavenge pump (described in *section 10.8*).

10.8 Lubrication

We saw in relation to the fuel tank that measures must be taken to prevent lateral cornering forces causing fuel surge. **Oil surge** is an even greater problem. Because motorcycles lean when cornering, the direction of the resultant *g* force is always down the centre of the engine and oil surge is not an issue. However, in a car during long corners, the oil pump pick-up point in

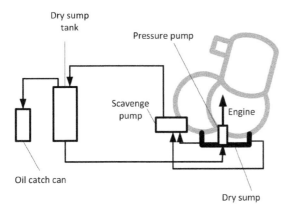

Figure 10.9
Dry sump system

the sump is likely to draw-in air as the oil moves to the side. A high revving race engine can be ruined if starved of oil for only a few seconds. It is likely to suffer failure of big-end bearings and subsequent damage to the crankshaft and connecting rods. There are three common approaches to alleviate the problem. In order of complexity, cost and effectiveness these are:

1. Introducing a thin horizontal **baffle plate** into the sump to reduce movement of the oil. This needs to fit closely around components but also to contain drain holes to allow oil to drip back into the sump. Also it may help to slightly overfill with oil – say an extra 0.5 litre.
2. Using an oil accumulator such as an 'Accusump'. This is a cylindrical tank with a spring-loaded piston that is connected to one of the main oil-ways in the engine. As oil pressure builds-up the piston is displaced and the cylinder holds a quantity of oil which is injected back into the engine if the oil pressure drops.
3. Adding a **dry sump**. A dry sump replaces the existing sump oil reservoir with an external tank which supplies the main oil pressure pump directly, usually by gravity feed. *Figure 10.9* shows the main components of the system. The original sump is replaced by a shallower dry sump. An additional **scavenge pump** sucks oil from each side of the dry sump and deposits it in the **dry sump tank**. The scavenge pump is often a two-stage gear pump and is driven either by the original water pump drive-shaft or a separate belt-driven pulley. The dry sump tank should spin the oil at the top to de-air it and be tall and narrow to reduce the effect of oil surge. It may be convenient to provide an oil-cooler in the line between the scavenge pump and the dry sump tank. A further important advantage of a dry sump is that it is generally shallower than the original sump and hence the engine's centre-of-mass can be lowered in the chassis.

SUMMARY OF KEY POINTS FROM CHAPTER 10

1. The normally aspirated four-stroke petrol engine draws an air/fuel mixture into the cylinder as the piston descends. The inlet valve closes and the piston rises to compress the fuel mixture. The spark plug fires to explode the fuel and force the piston down, generating power. The piston rises to expel gases through the exhaust valve.
2. Careful tuning of the lengths of inlet and exhaust pipes, together with valve overlap, can overcharge the cylinder with air which, when combined with additional fuel, can increase engine torque and power.
3. A restrictor in the air intake system provides a good maximum power limit as airflow starts to choke as it approaches the speed of sound.
4. Forcing additional air into the cylinders with either a supercharger or turbocharger can significantly increase power output.
5. Fuel tanks must be provided with a system to prevent fuel surge when cornering.
6. The engine management system controls the timing of spark plug ignition and the duration of fuel injection through look-up tables or maps.
7. Stand-alone engine management systems offer many features for the racer such as launch control and traction control.
8. The oil system must prevent oil starvation caused by oil surge when cornering. The most effective means of achieving this is with a dry sump system.

11 Set-up and testing

> **LEARNING OUTCOMES**
>
> At the end of this chapter:
> - You will understand what must be done to prepare a new car for racing
> - You will learn how to set up the suspension geometry including ride height, corner weights, camber and toe
> - You will understand the benefits to be gained from a data-logging system
> - You will learn how to carry out dynamic tests to optimise mechanical and aerodynamic balance of the car
> - You will learn how to set dampers on the track

11.1 Introduction

This chapter is concerned with getting a car into a state where it is fit to race. As Carroll Smith makes clear (*ref. 23*):

> *'The best designed and constructed chassis, suspension and tyres won't do you a bit of good unless they are precisely and correctly aligned.'*

Before the suspension set-up begins it is important to ensure the following:

1. The car should be complete including bodywork.
2. All fluids should be topped up to race levels.
3. All nuts and bolts, including wheel nuts, should be torqued up to race levels.
4. Wheel bearings should be adjusted to remove any play.
5. With the car off the ground, spring seats should be wound hand-tight with the damper fully extended (unless a specific spring pre-load has been imposed).
6. Adjustable dampers should be set to their softest setting.
7. Suspension links should be assembled so that plenty of adjustment is possible in both directions and both sides of the car are the same.

11.2 Suspension set-up

There is a wide spectrum of equipment available for wheel alignment, ranging from a simple straight-edge, string and builder's spirit level to expensive computer-based laser systems. At an intermediate level are commercially available camber and toe gauges. The following procedure assumes only the basic equipment and good results can be obtained, particularly if you are prepared to take an iterative approach and repeat the steps until all the parameters are fully achieved.

Set-up must be carried out with the wheels on a level surface. If such a surface is not available, pieces of sheet material should be used as shims to pack-up one wheel at each end of the car until level as indicated by a spirit level and straight-edge placed across the top of the tyres.

Step 1 – Tyre pressures. Check tyres are at the initial target pressures (although these may be modified after track testing).

Step 2 – Ride height. With inboard springs, ride height can be adjusted either by changing the length of push/pullrods or by turning the spring seats. The former is to be preferred as adjustment via the spring seats uses up valuable damper stroke or upsets the desired pre-load in the springs. Push/pullrods are normally constructed with a right-hand rose joint at one end and a left-hand rose joint at the other – both secured with lock-nuts. This means that the length of the rod, and hence the ride height, can be changed by rotating the rod. Spacer blocks should be prepared that give the specified ride height below easily accessible points at the front and rear of the car. At frequent intervals the car should be rolled forward and back a couple of metres and the suspension loaded in bump to alleviate the effect of arching from lateral wheel scrub.

Step 3 – Corner weights. It is important for good handling and grip that the static loads on the wheels are as equal as possible. It is however possible that one diagonal – say the front left wheel and the rear right wheel – support most of the static load of the car. To correct this we reduce the ride height of the 'strong' diagonal and raise the ride height of the 'weak' diagonal. The aim is to achieve equal loads on the two front wheels and equal loads on the two rear wheels. If the centre-of-mass of the car does not lie on the centre-line of the car, then, from simple equilibrium, it is not possible to equalise the wheel loads, but we can still equalise the sum of the two wheel loads on each diagonal – known as the **cross-weight**. All four corners should be adjusted in small increments so as to have the minimum effect on ride height. Special corner weight scales are best but for small motorcycle-engined cars it is possible to use cheap mechanical bathroom scales. The scales should be of identical thickness and placed on top of any ground-levelling shims. Calibrate them by standing on each in turn and adjusting until they give the same reading.

Chapter 11 **Set-up and testing**

In the following example we refer to *weight* in *kg* which is of course not correct however, by convention, *corner weights* are invariably output by scales in kg, and referred to as such in these units.

EXAMPLE 11.1

The following table gives initial corner weights (kg) for a small motorcycle-engined racing car. Calculate the optimum target corner weights.

	Left	Right
Front	60	77
Rear	90	80

Solution

Extending the table to give weight totals:

	Left	Right	Front/rear totals
Front	60	77	137
Rear	90	80	170
Left/right totals	150	157	Grand total = 307

Front percentage	= 100 × 137/307	=	44.6%
Rear percentage	= 100 − 44.6	=	55.4%
Left/front target	= 150 × 0.446	=	**67 kg**
Left/rear target	= 150 × 0.554	=	**83 kg**
Right/front target	= 157 × 0.446	=	**70 kg**
Right/rear target	= 157 × 0.554	=	**87 kg**

Check

Left/front − right/rear cross-weight	=	67 + 87	= 154 kg
Right/front − left/rear cross-weight	=	70 + 83	= 153 kg

Answer

	Left	Right
Front	67	70
Rear	83	87

Race car design

Step 4 – Camber. A common means of adjusting camber is by adding or removing small shims between the ends of the top wishbones and the uprights. This method has the advantage of being able to change camber without altering kingpin inclination. Camber is measured with a camber gauge which usually incorporates a spirit level bubble and some means of reading the angle from vertical. Although not as quick and easy, good results can be obtained from a simple builder's spirit level (or even a plumb line) by measuring the vertical (h) and horizontal (l) distance between two points at the top and bottom of the wheel rim:

Camber angle = $\tan^{-1}(l/h)$

It clearly helps to be aware of the precise change in camber caused by the addition or removal of specific shim thicknesses. Camber within 0.25° is sufficiently accurate.

Step 5 – Toe. Few things improve both the appearance and performance of a car more than getting the wheels pointing in the right direction! This needs to be carried out with care and precision. The simple method described here involves the construction of an external reference frame consisting of two lines parallel to the longitudinal centre-line of the car – *Figure 11.1* shows the arrangement.

Two lengths of fishing line are connected a precise distance apart from two steel tubes. The gap between the lines should be the overall width of the car plus say 150 mm. The tubes are supported on axle stands at each end of the car so that the fishing lines are stretched taught at the centre of wheel height. The tubes are moved carefully from side-to-side until the gaps between the lines and the wheel hub centres are the same on both sides of the car. Careful measurements should be taken with an engineer's steel rule. The front distances are likely to be different from the rears because of varying front and rear track dimensions.

The process starts by centring the steering rack, i.e. ensuring that equal lengths of rack project from each side of the rack housing. The rack should now be locked in place. This can be by half-tubes of a precise length that drop

Figure 11.1
Setting static toe

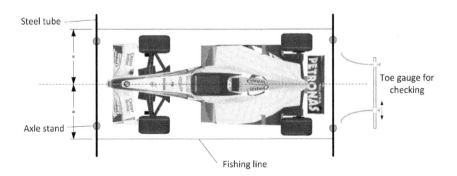

on the rack projections to prevent movement or simply by adding Jubilee clips to the rack. If not already done, the steering column is now moved on the splines until the steering wheel is in the straight-ahead position. The distance to the front and rear face of each wheel rim is now measured and adjusted, via the toe-control rod, to obtain the desired toe setting. Once complete, this can be checked with a toe gauge – an example of which is shown in *Figure 11.1*. The gauge is adjusted to touch the insides of the wheel rims at the front and rear of the wheels. The toe control rods are locked with nuts and the steering rack unlocked.

Step 6 – Dampers. *Examples 4.4* and *4.5* showed how to calculate the initial damping coefficients for a car. For adjustable dampers, these values should be compared to the slopes of the damper characteristic curves obtained from tests at various hardness settings. The aim is to determine the initial settings that best match the calculated damping coefficients.

11.3 Testing

Before testing, the newly completed car must be subjected to a rigorous check on the tightness of nuts, bolts and hose clamps together with fluid levels. It is useful if the car is securely raised off the ground on robust stands and the largest member of the team applies significant force to each wheel in all directions. Other than on the steering axis, there should be no relative movement between the wheels and the chassis.

Dynamic testing should ideally be carried out by an experienced driver who can provide good feedback to the design team on issues such as understeer/oversteer balance.

11.3.1 Safety

At racing events the circuit regulations, scrutineers and marshals enforce a relatively safe operating environment, however during informal testing it is often the team which must take responsibility for safe procedures. The testing session should be the subject of a written formal risk assessment which is read and understood by all and signed off by senior management. Some issues which should be covered include the following:

Transport and loading/unloading

- The car must be properly secured for transport
- The driver must be qualified and trained to drive the vehicle
- An adequate number of team members must be available for loading/unloading and no member expected to lift more than an agreed weight.

Race car design

Driving

- Drivers must wear full protective gear and a safety harness whenever the car is under power
- All drivers must have previously demonstrated the ability to exit the car within five seconds (Including removing safety harness)
- The test track must contain adequate run-off areas at all corners
- The test route must be well clear of other vehicles and any other hard objects.

Spectating

- When the car is running, spectators must stay within a designated area well outside any run-off area.

Fire and fuel

- Two designated team members should be equipped with approved fire extinguishers at all times
- Fuel must be stored in approved containers
- The driver must remain outside the car during refuelling.

11.3.2 Data acquisition

The value of testing is considerably enhanced if key data can be logged for subsequent analysis. A data-acquisition system has become an indispensable component for effective testing and development of both car and driver. Stand-alone engine management systems usually have the ability to log engine-related parameters such as engine rpm, water temperature, oil pressure and exhaust oxygen (lambda) – provided appropriate sensors are fitted of course. However dedicated data loggers, often associated with cockpit instruments, can provide much more, particularly if linked to inertial navigation or satellite GPS systems. This enables monitoring of speed, lateral and longitudinal g force etc., which can be directly related to lap position on a generated map.

Data loggers can generally be upgraded to provide additional analogue channels which can be used to provide the designer with a wide range of valuable information. Strain gauges can be added to give structural loads in the chassis and suspension links, rotary potentiometers can provide steering angle, tilt sensors can give roll and pitch angles and proximity sensors can output ground clearance. Of particular value are linear potentiometers to record movement of the spring/dampers or bell-cranks. Not only does this provide data on suspension movement during acceleration, braking and cornering, but vital information on front and rear downforce can also be extracted to help with achieving the aerodynamic balance of the car.

Data is logged at regular intervals and the sampling rate can often be set by the user to suit the data of interest. Generally good results can be obtained if sampling is at twice the frequency of the event. For logging events on the circuit that change relatively slowly, such as speed, engine revs, oil pressure and water temperature, a rate of 10 Hz is sufficient. We saw in *Chapter 4* that the typical frequency of an unsprung wheel mass is about 20 Hz, so if you wish to investigate detailed wheel movements a sampling rate of at least 40 Hz is required. An engine at 12 000 rpm is rotating at 200 Hz, so 400 Hz is necessary for detailed engine events. Of course, higher sampling rates generate data and use up memory much more quickly. Software tools are provided for interpretation and analysis of the large amounts of data generated.

11.3.3 Warm-up test

The engine is run until all the fluids are up to working temperature and pressure. All pipework is inspected carefully for leaks. The functioning of sensors and instruments is checked where possible. A sound test can be carried out.

11.3.4 Shakedown test

The car is run at progressively quicker speeds with frequent pit-stops to check for loose nuts and bolts and leaks. Oil pressure and water temperature should be carefully monitored and logged for future analysis.

11.3.5 Tyre temperature test

The car is run at racing speeds until the tyres are up to temperature. While still hot, a pyrometer is used on each tyre to determine the temperature say 25 mm from the inside edge, the middle and 25 mm from the outside edge of the tread. A probe-type pyrometer is best which should be inserted at a constant 3 mm into the tread. The heavily loaded tyre, i.e. on the left-hand-side of the car on a clockwise circuit, should register temperatures in the 80°C–105°C range. The inside edge should read about 5°C more than the outside. *Table 11.1* indicates how the three temperatures are interpreted.

Table 11.1 Tyre temperature testing

Symptom	Interpretation
Inner edge more than 5°C hotter than outer	Too much negative camber
Inner edge less than 5°C hotter than outer	Not enough negative camber
Centre greater than inner edge	Too much tyre pressure
Centre less than edge average	Too little tyre pressure
Fronts hotter than rears	Understeering car
Rears hotter than fronts	Oversteering car

11.3.6 Mechanical balance

The mechanical undertsteer/oversteer balance test is carried out at relatively slow speeds before aerodynamic downforce has a chance to build. Nevertheless it makes sense to minimise wing angles of attack. Ideally the test is carried out on a circular skid pad of minimum radius 30 m. The car is required to follow a constant radius line at gradually increasing speed until a slide occurs. If the steering angle is plotted against speed (or lateral *g*) a graph resembling the handling curve shown in *Figure 5.16* will result. The final slide will result in either the car leaving the circle and carrying straight-on (understeer) or entering a spin (oversteer). For cars with a high power-to-weight ratio it is important that inexperienced drivers do not confuse mechanical oversteer with power oversteer. Applying too much throttle at any time can cause the rear of a powerful car to break-away and spin as lateral traction is reduced in the presence of longitudinal grip.

Table 5.3 listed the measures that can be taken to adjust the balance of the car.

It should be noted that constant turning in one direction at high lateral acceleration is the primary cause of oil surge and hence it is important to monitor oil pressure carefully during this test.

11.3.7 Aerodynamic balance

We know that aerodynamic downforce is proportional to the speed squared and hence this test is carried out on fast corners – with plenty of run-off in case of a slide. The aim is to adjust the angles of attack of the front and rear wings to achieve the desired understeer/oversteer balance. To reduce oversteer, either reduce front wing downforce or increase the rear. Is it also worth checking that the overall downforce levels, and hence drag, are appropriate for achieving the maximum speed on the straights.

11.3.8 Damper settings

Examples 4.4 and *4.5* showed how to calculate initial settings for adjustable dampers. It may be desirable to modify these settings to accommodate specific track conditions or the preferences of particular drivers. The time-dependent load transfer characteristics of dampers should not be used to establish the steady-state balance of a car but adjustments do affect transient balance such as during corner-entry and corner-exit. Neither should dampers be used to limit excessive corner roll. The following procedure is extracted from Milliken and Milliken (*ref. 15*) and is based on recommendations from the Koni damper company. It assumes at least two-way adjustable dampers (bump and rebound).

1. **Setting bump damping to control unsprung wheel assembly oscillation**
   (a) Set all dampers to minimum bump and rebound and drive a lap or two. Note how the car responds on bumpy corners. The wheels are likely to bounce or 'side-hop' as they are inadequately damped.
   (b) Increase bump adjustment by three clicks and repeat, noting any improvement.
   (c) Repeat (b) until the car feels hard over bumps.
   (d) Back off bump adjustment by two clicks.

2. **Setting rebound damping to control transient corner roll**
   (a) Leave bump damping as above and drive a lap or two. Note how the car rolls when entering a corner.
   (b) Increase rebound damping by three clicks on all wheels and repeat laps.
   (c) Repeat (b) until the car enters the turns smoothly with little roll on initial turn-in.
   (d) Increase rear rebound damping and/or reduce front rebound damping to reduce transient corner-entry understeer and vice versa.
   (e) Reduce rebound damping if the car tends to 'jack down' on repeated bumps.

Having completed the above, it would clearly be instructive to compare the damper settings to the initial calculated values.

> **SUMMARY OF KEY POINTS FROM CHAPTER 11**
> 1. The suspension should be set up with care and precision and includes tyre pressures, ride height, corner weights, wheel camber and toe.
> 2. The team must follow agreed safety procedures during testing.
> 3. The addition of a data-logging system greatly enhances the value of dynamic testing and development.
> 4. A tyre temperature test provides a good check on the suitability of tyre pressures and wheel camber.
> 5. The mechanical understeer/oversteer balance of a car is established at relatively slow speeds, ideally on a circular skid pad.
> 6. The aerodynamic balance of a car is established at faster speeds and involves setting the angles of attack of the front and rear wings.
> 7. Initial calculated damper settings can be modified on track to suit specific conditions and driver preferences.

Appendix 1

Deriving Pacejka tyre coefficients

We stated in *Chapter 5* that there are advantages in having tyre data presented in the form of a mathematical model so that, among other things, understeer/oversteer balance calculations can be computerised. This appendix shows how the key Pacejka coefficients can be obtained from graphical tyre test results and goes on to explain how the values are modified to represent tyre variations such as different tread widths or softer compounds.

We will only consider the *lateral grip* form of the Pacejka 'magic' formula as detailed in *section 5.2.6*. Once the formula is populated with the appropriate coefficients from actual tyre tests the designer can simply **input**:

- vertical tyre load, slip angle and camber angle

and the **output** is:

- lateral cornering or grip force.

The following procedure is based on *ref. 17*. All angles are in radians. The first point to make is that some Pacejka coefficients are more significant than others. Of the eighteen coefficients listed in *Table 5.1* about five can be simply set to zero without unduly affecting the performance of the formula in most cases. These are shown struck-out in *Table A1.1*.

Consider the test curves shown in *Figure 5.8a* for Avon British F3 tyre at 0° camber and reproduced in *Figure A1.1*. This is a 180/550 R13 radial tyre. This particular set of curves is for the zero camber case and this will be used to derive the basic formula coefficients. Other graphs exist for non-zero camber and an example is shown in *Figure A1.4* on page 256. Our objective is a formula that can reproduce these curves as well as those at non-zero camber values. The dashed lines in *Figure A1.1* have been added to show the curve offsets from the origin.

Appendix 1   Deriving Pacejka tyre coefficients

Table A1.1   Pacejka coefficients

	Description
F_{Z0}	Nominal load (N)
p_{CY1}	Shape factor
p_{DY1}	Lateral friction, μ_y
p_{DY2}	Variation of friction with load
p_{DY3}	Variation of friction with camber squared
p_{EY1}	Lateral curvature at F_{Z0}
~~p_{EY2}~~	~~Variation of curvature with load~~
~~p_{EY3}~~	~~Zero order camber dependency of curvature~~
~~p_{EY4}~~	~~Variation of curvature with camber~~
p_{KY1}	Maximum value of stiffness K_y/F_{Z0}
p_{KY2}	Normalised load at which K_y reaches max. value
p_{KY3}	Variation of K_y/F_{Z0} with camber
p_{HY1}	Horizontal shift S_{Hy} at F_{Z0}
~~p_{HY2}~~	~~Variation of S_{Hy} with load~~
p_{HY3}	Variation of S_{Hy} with camber
p_{VY1}	Vertical shift S_{Vy} at F_{Z0}
~~p_{VY2}~~	~~Variation of S_{Vy} with load~~
p_{VY3}	Variation of S_{Vy} with camber
p_{VY4}	Variation of S_{Vy} with camber and load

Step 1 – Nominal load, F_{Z0}

The first stage is to specify the nominal load, F_{Z0}, on which the fundamental curve will be based. It makes sense to choose one of the load increments in the test data. Also the nominal load should be close to the maximum vertical cornering load of the car. For small single-seater cars this is likely to be around 200 kg – hence choose this value. Also start with the zero camber curve.

Nominal load, F_{Z0} = 200 × 9.81 = **1962 N**

Step 2 – Friction coefficient parameter at nominal load, p_{DY1}

The peak value of cornering force, D_y, is given by *equation [5.6]*:

$$D_y = F_Z(p_{DY1} + p_{DY2}df_Z)(1 - p_{DY3}\gamma_y^2)\lambda_{\mu y}$$

At the nominal load ($F_Z = F_{Z0}$) zero camber and $\lambda_{\mu y} = 1$, this simplifies to:

$$D_y = F_{Z0} \times p_{DY1}$$

Appendix 1 Deriving Pacejka tyre coefficients

Figure A1.1
Cornering force for Avon British F3 tyre at zero camber (reproduced with kind permission from Avon Tyres Motorsport)

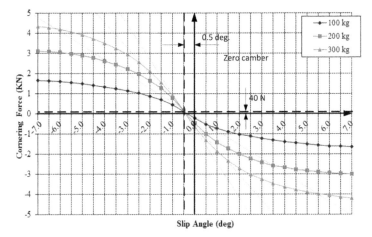

From the graph in *Figure A1.1* it can be seen that the 200 kg curve is very close to reaching a peak value at 7° of slip, but it would be better if the testing had continued to higher slip angles, ideally beyond the peak. Also note that the above curves are not quite symmetrical, with a small vertical displacement at the origin, leading to slightly different cornering force values, F_y, at +7° and −7°. This is likely to be caused by asymmetry in tyre construction. The initial base curve is assumed to pass through the origin, hence take the average value of cornering forces from above and below the horizontal axis:

$$D_y = 3050 \text{ N}$$

Friction coefficient, $\mathbf{p_{DY1}} = D_y/F_{Z0}$ = 3050/1962 = **1.55**

Step 3 – Stiffness parameters, $\mathbf{p_{KY1}}$ and $\mathbf{p_{KY2}}$

Now turning to cornering stiffness – i.e. the slope of the curves at the origin. Unusually these parameters are based on the maximum stiffness case and not the nominal load case. It can be seen in *Figure A1.1* that dashed lines have been drawn parallel to the curves at the origin. Again taking average values for positive and negative slip angles and converting to radians:

At 100 kg

Slope of line, k_{y100} = 5000/(−6.3 × π /180) = −40 340 N/rad

Normalise by dividing by F_{Z0} = −40 340/1962 = **−20.6**

At 200 kg

Slope of line, k_{y200} = 5000/(−3.9 × π /180) = −73 456 N/rad

Normalise by dividing by F_{Z0} = −73 456/1962 = **−37.4**

At 300 kg

Slope of line, k_{y300} = 5000/(−2.95 × π /180) = −97 111 N/rad

Normalise by dividing by F_{Z0} = −97 111/1962 = **−49.5**

$\mathbf{p_{KY1}}$ is the maximum value of the stiffness/F_{Z0}. To obtain this it is neces-

Appendix 1 Deriving Pacejka tyre coefficients

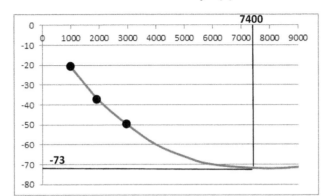

Figure A1.2
Maximum value of stiffness

sary to plot the above stiffness values and estimate where the resulting curve peaks, which, from *Figure A1.2*, is at about **−73.0**.

p_{KY2} is defined as the 'normalised load at which K_y reaches maximum value'. This is estimated below at 7400 N, hence we get:

$$p_{KY2} = 7400/F_{Z0} = 7400/1962 = \mathbf{3.77}$$

Step 4 – Shape parameter, p_{CY1}

This parameter determines the degree to which the peak cornering force falls off at high slip angles. It is given by the formula:

Ref. 11 $\quad p_{CY1} = 1 + \left(1 - \dfrac{2}{\pi} \sin^{-1} \dfrac{y_a}{D_y}\right)$

where $\quad y_a$ = distance to horizontal asymptote of the F_{Z0} curve

D_y = peak value of the F_{Z0} curve
(see *Figure A1.3*).

Values will generally lie in the range 1.1 to 1.6 with say 1.3 being typical for cross-ply tyres and around 1.5 for more 'peaky' radials. *Table A1.2* gives an indication of the degree to which the peak value reduces.

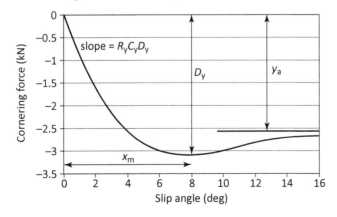

Figure A1.3
Deriving the curvature parameter p_{EY1}

Appendix 1 Deriving Pacejka tyre coefficients

Table A1.2 Degree to which peak values of p_{CY1} reduce

Value of p_{CY1}	% reduction on peak value
1.0	0 (i.e. curve levels off)
1.1	1
1.2	5
1.3	11
1.4	19
1.5	29

It is rare for test data to extend to sufficiently high slip angles for reliable estimates of p_{CY1} to be made, so the use of the typical values given above is suggested. In our case the tyre is a radial so use a value of **1.5**.

From *equation [5.8]*

$$C_y = p_{CY1}\lambda_{Cy} = p_{CY1} \quad \text{when } \lambda_{Cy} = 1.0$$

Step 5 – Curvature parameter, p_{EY1}

The curvature factor, E_y, controls the horizontal location of the peak, x_m, as shown in *Figure A1.3*. It is given by *equation [5.12]*:

$$E_y = (p_{EY1} + p_{EY2}df_Z)\{1 - (p_{EY3} + p_{EY4}\gamma_y)\text{sgn}(\alpha_y)\}\gamma_{Ey}$$

Simplifying at the nominal load ($F_Z = F_{Z0}$), zero camber, $p_{EY3} = 0$ and $\lambda_{Ey} = 1.0$:

$$E_y = p_{EY1}$$

Ref. 17
$$E_y = p_{EY1} = \frac{B_y x_m - \tan(\pi/2p_{CY1})}{B_y x_m - \tan^{-1}(B_y x_m)}$$

From *Figure A1.1* we can estimate that the curve will peak at about 8° slip angle:

$$x_m = 8° = 0.14 \text{ rad}$$

Slope of F_{Z0} curve at origin $= -73\,456 \text{ N/rad} = B_y C_y D_y$

Hence
$$B_y = -73\,456/C_y D_y$$
$$= -73\,456/(1.5 \times -3050)$$
$$= 16.1$$

$$\therefore B_y x_m = 16.1 \times 0.14 = 2.24$$

$$p_{EY1} = \frac{2.24 - \tan(\pi/(2 \times 1.5))}{2.24 - \tan^{-1}(2.24)}$$

$$= (2.24 - 1.73)/(2.24 - 1.15)$$

$$p_{EY1} = 0.47$$

Appendix 1 Deriving Pacejka tyre coefficients

Step 6 – Horizontal shift at F_{Z0}, parameter p_{HY1}

The dashed lines in *Figure A1.1* indicate the curve offsets from the origin. The units of horizontal shift are radians.

Estimate
$$\text{Horizontal shift } p_{HY1} = 0.50° \times \pi/180 = 0.0087$$

Notice that the sign is positive as shift to the left increases the effective slip angle.

Step 7 – Vertical shift at F_{Z0}, parameter p_{VY1}

The dashed lines in *Figure A1.1* indicate the curve offsets from the origin. The units of vertical shift are force in Newtons which is then normalised by dividing by F_{Z0}:

Estimate
$$\text{Vertical shift } p_{VY1} = 40/1962 = 0.02$$

Variation with vertical load, F_Z

Step 8 – Friction coefficient at other vertical loads, parameter p_{DY2}

We now need to know how the friction coefficient varies with load:

$$\text{At 300 kg load, } F_Z = 300 \times 9.8 = 2943 \text{ N}$$
$$\text{Friction coefficient, } \mu = F_y/F_Z = 4.30/2.943 = 1.46$$

Comparing this with the friction coefficient at the nominal load, F_{Z0}:

$$\Delta F_Z = 1962 - 2943 = -981$$
$$\Delta \mu = 1.55 - 1.46 = 0.09$$

Nominal change in friction with load, $\mathbf{p_{DY2}}$
$$= 0.09/-981 \times 1962 = \mathbf{-0.18}$$

Variation with camber, γ

Step 9 – Friction coefficient at other wheel cambers, parameter p_{DY3}

Figure A1.4 shows the tyre test curve for 4.0° of camber. It can be seen that the average peak value from above and below the horizontal axis is:

$$D_y = 3.10 \text{ kN (3.05 kN at zero camber)}$$

Again from *equation [5.6]*
$$D_y = F_Z(p_{DY1} + p_{DY2}df_Z)(1 - p_{DY3}\gamma_y^2)\lambda_{\mu y}$$

At the nominal load ($F_Z = F_{Z0}$) and $\lambda_{\mu y} = 1$, this simplifies to:

Appendix 1   Deriving Pacejka tyre coefficients

Figure A1.4
Cornering force for Avon British F3 tyre at 4.0° camber (reproduced and adapted with kind permission from Avon Tyres Motorsport)

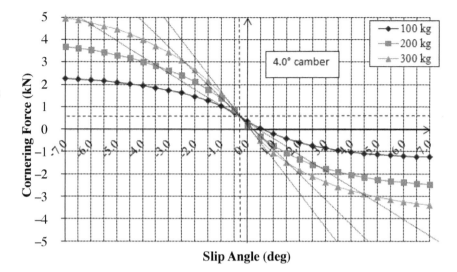

$$D_y = F_{Z0}p_{DY1}(1 - p_{DY3}\gamma_y^2) = 3100$$

$$\gamma_y = 4.0° = 4 \times \pi/180 = 0.0698 \text{ rad}$$

$$\therefore 1962 \times 1.55(1 - p_{DY3} \times 0.0698^2) = 3100$$

$$p_{DY3} = -3.98$$

Step 10 – Variation of horizontal shift with wheel camber, parameter p_{HY3}

It can be seen from *Figures A1.1* and *A1.4* that the horizontal shift of the dashed line has moved from 0.50° to about 0.35°, i.e. −0.15° = −0.0026 rad.

From *equation [5.11]*:

$$S_{Hy} = (p_{HY1} + p_{HY2}df_Z + p_{HY3}\gamma_y)\lambda_{Hy}$$

It can be seen from the above that the horizontal shift due to camber of 0.0026 rad must be caused by the $p_{HY3}\gamma_y$ term.

$$\therefore p_{HY3} \times 0.0698 = -0.0026$$

$$p_{HY3} = -0.037$$

Step 11 – Variation of vertical shift with wheel camber, parameter p_{VY3}

It can be seen from *Figures A1.1* and *A1.4* that the vertical shift of the dashed line has moved from + 40 N to about + 600 N, i.e. + 560 N.

From *equation [5.13]*:

$$S_{Vy} = F_Z\{p_{VY1} + p_{VY2}df_Z + (p_{VY3} + p_{VY4}\,df_Z)\gamma_y\}\lambda_{Vy}\lambda_{Kya}$$

It can be seen from the above that the vertical shift of 560 N must be caused by the $p_{VY3}\gamma_y$ term.

$$\therefore 1962 \times p_{VY3} \times 0.0698 = 560$$

$$p_{VY3} = 4.09$$

Step 12 – Variation of vertical shift with wheel camber and load, parameter p_{VY4}

It can be seen in *Figure A1.4* that the vertical shift is identical for all the curves. A problem is that *equation [5.13]* above indicates that the magnitude of the vertical shift caused by p_{VY3} will increase with F_Z. We must therefore use the p_{VY4} parameter to counteract this effect. We essentially need the vertical shift to stay constant at about 560 N for 4°camber. If we consider the F_Z value of 2943 N (300 kg):

Increase in vertical shift = $(2943/1962 \times 560) - 560$ = 280 N

We therefore need the p_{VY4} parameter to generate −280 N of vertical shift:

From *equation [5.13]*:

$$F_Z\, p_{VY4}\, df_Z \gamma_y = -280$$

where $\quad df_Z = (2943 - 1962)/1962 \quad = 0.5$

$$\therefore \quad p_{VY4} = -280/(2942 \times 0.5 \times 0.0698)$$

$$p_{VY4} = -2.73$$

Step 13 – Variation of maximum value of stiffness K_y/F_{Z0} with wheel camber, parameter p_{KY3}

The maximum value of the cornering stiffness reduces with the absolute value of wheel camber. Therefore repeat *step 3* to obtain the maximum slopes of the curves at 4° camber as shown in *Figure A1.4*. This is shown in *Figure A1.5* where the result at 4° is compared to the zero camber case.

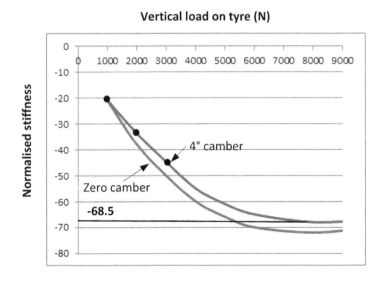

Figure A1.5
Maximum value of stiffness at 4.0° camber

Appendix 1 Deriving Pacejka tyre coefficients

At 100 kg

$$\text{Slope of line, } k_{y100} = 5000/(-7.1 \times \pi/180) = -40\,349 \text{ N/rad}$$

$$\text{Normalise by dividing by } F_{Z0} = -41\,518/1962 = \mathbf{-20.6}$$

At 200 kg

$$\text{Slope of line, } k_{y200} = 5000/(-4.4 \times \pi/180) = -65\,109 \text{ N/rad}$$

$$\text{Normalise by dividing by } F_{Z0} = -73\,456/1962 = \mathbf{-33.2}$$

At 300 kg

$$\text{Slope of line, } k_{y300} = 5000/(-3.3 \times \pi/180) = -86\,812 \text{ N/rad}$$

$$\text{Normalise by dividing by } F_{Z0} = -97\,111/1962 = \mathbf{-44.2}$$

It can be seen that the normalised maximum stiffness value changes from −73 to −68.5.

$$\text{Maximum stiffness at zero camber} = 73 \times 1962 = 145\,200 \text{ N/rad}$$

$$\text{Maximum stiffness at 4° camber} = 68.5 \times 1962 = 134\,400 \text{ N/rad}$$

Ref. 17
$$134\,400 = 145\,200(1 - p_{KY3}|\lambda|)$$

$$p_{KY3} = 10\,800/(145\,200 \times 0.0698)$$

$$\mathbf{p_{KY3} = 1.07}$$

Summary and conclusions

Table A1.3 shows the hand-derived values of Pacejka parameters compared to the original computer-generated values from Avon. It is clear that there are substantial differences. Some discrepancies can be explained by the adoption of different nominal loads.

Figure A1.6
Pacejka versus test data for 300 kg load and 3° camber

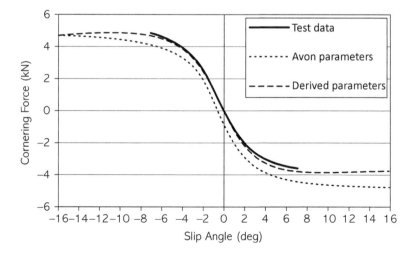

Appendix 1 Deriving Pacejka tyre coefficients

Table A1.3 Pacejka coefficients – Avon British F3 tyres

	Avon	Derived	Description
$F_{Z0} =$	2444	1962	Nominal load (N)
$p_{CY1} =$	0.324013	1.5	Shape factor
$p_{DY1} =$	−3.674945	−1.55	Lateral friction, μ_y
$p_{DY2} =$	0.285134	−0.18	Variation of friction with load
$p_{DY3} =$	−2.494252	−3.98	Variation of friction with camber squared
$p_{EY1} =$	−0.078785	0.47	Lateral curvature at F_{Z0}
$p_{EY2} =$	0.245086	0	Variation of curvature with load
$p_{EY3} =$	−0.382274	0	Zero order camber dependency of curvature
$p_{EY4} =$	−6.25570332	0	Variation of curvature with camber
$p_{KY1} =$	−41.7228113	−73.0	Maximum value of stiffness K_y/F_{Z0}
$p_{KY2} =$	2.11293838	3.77	Normalised load at which K_y reaches max. value
$p_{KY3} =$	0.150080764	1.07	Variation of K_y/F_{Z0} with camber
$p_{HY1} =$	0.00711	0.0087	Horizontal shift S_{Hy} at F_{Z0}
$p_{HY2} =$	−0.000509	0	Variation of S_{Hy} with load
$p_{HY3} =$	0.049069131	0.037	Variation of S_{Hy} with camber
$p_{VY1} =$	−0.00734	0.02	Vertical shift S_{Vy} at F_{Z0}
$p_{VY2} =$	−0.0778	0	Variation of S_{Vy} with load
$p_{VY3} =$	−0.0641	4.09	Variation of S_{Vy} with camber
$p_{VY4} =$	−0.6978041	−2.73	Variation of S_{Vy} with camber and load

The ultimate test of the parameter values is how well they reproduce the test data. *Figure A1.6* shows the above derived parameters compared to the test data for a case that was not directly considered in the derivation, i.e. 300 kg vertical load at 3°camber.

Also shown is the curve based on the Avon Pacejka parameters. It can be seen that the derived data fits well, particularly in the negative camber sector where it is virtually indistinguishable from the test data. The Pacejka curves have been extended to ±16° and it can be seen that the derived data eventually peaks at about ±10°, whereas the Avon curve continues to increase. In both cases the curve for a radial tyre is perhaps expected to be more 'peaky'. The Avon curve would benefit from an increase in positive shift with camber, i.e. a bigger value for p_{VY3}. However, apart from this, it is somewhat surprising how similar the two Pacejka curves are, given the significant variation in parameter values.

Appendix 1   Deriving Pacejka tyre coefficients

Note on signs

It is common to present Pacejka curves with the opposite slope to the Avon curves presented above – i.e. starting in the bottom left segment and ending in the top right. Pacejka himself does this. The main required changes to the parameters in order to achieve this involve reversing the sign of the friction coefficient, p_{DY1}, and ensuring that the stiffness parameters, p_{KY1} and p_{KY2}, have the same sign.

Parameter modifications for tyre, wheel and road surface variations

Different friction surfaces

It is common for tyre test rigs to over-estimate the friction level available on real roads, particularly for very soft tyres. Where track testing reveals a discrepancy it is a simple matter to set the Pacejka scaling factor, $\lambda_{\mu y}$, to a value less than one in *equation [5.6]* such that the new peak cornering force corresponds to a realistic values obtained on the track. This modifies the peak cornering force without unduly affecting the shape or slope of the curve.

Different tyre compounds

Tyre manufacturers produce a 'family' of tyres with the same basic construction but with compounds of varying hardness and with different section widths. If test data is available for one member of the family, it is possible to estimate parameters to enable modelling of related tyres to be done.

In general softer compounds will produce more grip, however manufacturers tend not to provide meaningful hardness data on race compounds. The friction scaling factor $\lambda_{\mu y}$ can however be increased for a softer compound.

Different tyre widths

A wider tyre has higher stiffness and provides more peak grip. Dixon (*ref. 7*) suggests that cornering stiffness varies with width to the power 0.3, although this is reduced by about half if the pressure of a wider tyre is reduced to maintain a constant contact pressure. He also suggests that maximum cornering force varies with width to the power of 0.15, provided tyre pressures are adjusted to keep the contact patch length constant.

The Pacejka scaling factors λ_{Ky} and $\lambda_{\mu y}$ can therefore be adjusted as follows if a tyre is increased from say 180 mm width to 250 mm:

$$\lambda_{Ky} = (250/180)^{0.3} = 1.10 \text{ (1.05 with reduced tyre pressure)}$$

$$\lambda_{\mu y} = (250/180)^{0.15} = 1.05$$

The above approach can also be used to predict the reduction in dry grip when a slick tyre is grooved to provide drainage channels in the wet.

Appendix 1 Deriving Pacejka tyre coefficients

Different wheel widths

For each tyre size, manufacturers will generally provide guidance on the range of wheel rim widths that are suitable. Dixon (*ref. 7*) suggests that cornering stiffness varies with rim width to the power of 0.5. Thus changing from a 200 mm rim to a 225 mm rim produces a change in λ_{Ky} as follows:

$$\lambda_{Ky} = (225/200)^{0.5} = 1.06$$

Different tyre pressure or temperature

Unlike some tyre models, Pacejka does not contain specific parameters for modelling changes in tyre pressure or temperature, although the model can be extended to do so. It has been found that the optimum pressure for peak grip increases as the vertical load increases. Puhn (*ref. 21*) provides graphs showing maximum cornering force reducing by up to 20% if a tyre is not used at its optimum pressure or temperature. The friction scaling factor $\lambda_{\mu y}$ can be set to <1.0 to compensate for such situations.

Appendix 2

Tube properties

Imperial steel tube properties (see note on page 41)

Diameter, D (mm)	Thickness, t (mm)	Area, A (mm^2)	Weight, w (kg/m)	2nd mom. area, I (mm^4)	Elas. mod., Z (mm^3)
28.58	1.22	118.0	0.925	13 992.9	874.6
	1.63	138.0	1.082	12 575.0	880.0
	2.03	191.1	1.498	21 557.8	1347.4
25.40	1.22	92.7	0.727	6790.4	534.7
	1.63	121.7	0.954	8637.2	680.1
	2.03	149.0	1.168	10 251.7	807.2
22.22	1.22	80.5	0.631	4451.9	400.7
	1.63	105.4	0.827	5622.5	506.1
	2.03	128.8	1.009	6627.2	596.5
19.05	1.22	68.3	0.536	2728.4	286.4
	1.63	89.2	0.699	3413.3	358.4
	2.03	108.5	0.851	3986.3	418.5
15.88	1.22	56.2	0.441	1519.9	191.4
	1.63	73.0	0.572	1876.5	236.3
	2.03	88.3	0.692	2163.4	272.5
15.00	1.22	52.8	0.414	1263.5	168.5
	1.63	68.5	0.537	1552.6	207.0
	2.03	82.7	0.648	1781.9	237.6
12.70	1.22	44.0	0.345	733.0	115.4
	1.63	56.7	0.444	887.2	139.7
	2.03	68.0	0.533	1003.4	158.0

Appendix 2 **Tube properties**

Imperial square box properties

Side (mm)	Thickness, t (mm)	Area, A (mm²)	Weight, w (kg/m)	2nd mom. area, I (mm⁴)	Elas. mod., Z (mm³)
25.40	1.22	118.0	0.925	11 527.7	907.7
	1.63	155.0	1.215	14 662.9	1154.6
	2.03	189.8	1.488	17 403.9	1370.4
22.22	1.22	102.5	0.803	7557.7	680.3
	1.63	134.2	1.052	9545.1	859.1
	2.03	163.9	1.285	11 250.8	1012.7
19.05	1.22	87.0	0.682	4631.8	486.3
	1.63	113.6	0.890	5794.6	608.4
	2.03	138.2	1.084	6767.3	710.5
15.88	1.22	71.5	0.561	2580.3	325.0
	1.63	92.9	0.728	3185.6	401.2
	2.03	112.5	0.882	3672.7	462.6
15.00	1.22	67.2	0.527	2144.9	286.0
	1.63	87.2	0.683	2635.7	351.4
	2.03	105.3	0.826	3025.1	403.3
12.70	3.03	117.2	0.919	2005.9	315.9
	4.03	139.8	1.096	2129.2	335.3
	5.03	154.3	1.210	2163.8	340.8

Metric steel tube properties

Diameter, D (mm)	Thickness, t (mm)	Area, A (mm²)	Weight, w (kg/m)	2nd mom. area, I (mm⁴)	Elas. mod., Z (mm³)
30.00	1.50	134.3	1.053	13 673.7	911.6
	2.00	175.9	1.379	17 329.0	1155.3
	2.50	216.0	1.693	20 586.0	1372.4
25.00	1.50	110.7	0.868	7675.7	614.1
	2.00	144.5	1.133	9628.2	770.3
	2.50	176.7	1.385	11 320.8	905.7
22.00	1.50	96.6	0.757	5101.9	463.8
	2.00	125.7	0.985	6346.0	576.9
	2.50	153.2	1.201	7399.2	672.7
20.00	1.50	87.2	0.683	3754.2	375.4
	2.00	113.1	0.887	4637.0	463.7
	2.50	137.4	1.078	5368.9	536.9
18.00	1.50	77.8	0.610	2667.9	296.4
	2.00	100.5	0.788	3267.3	363.0
	2.50	121.7	0.954	3751.0	416.8
15.00	1.50	63.6	0.499	1467.2	195.6
	2.00	81.7	0.640	1766.4	235.5
	2.50	98.2	0.770	1994.2	265.9
12.00	1.50	49.5	0.388	695.8	116.0
	2.00	62.8	0.493	816.8	136.1
	2.50	74.6	0.585	900.0	150.0

Appendix 2 Tube properties

Metric square box properties

Side (mm)	Thickness, t (mm)	Area, A (mm²)	Weight, w (kg/m)	2nd mom area, I (mm⁴)	Elas. mod., Z (mm³)
25.00	1.50	141.0	1.105	13 030.8	1042.5
	2.00	184.0	1.443	16 345.3	1307.6
	2.50	225.0	1.764	19 218.8	1537.5
22.00	1.50	123.0	0.964	8661.3	787.4
	2.00	160.0	1.254	10 773.3	979.4
	2.50	195.0	1.529	12 561.3	1141.9
20.00	1.50	111.0	0.870	6373.3	637.3
	2.00	144.0	1.129	7872.0	787.2
	2.50	175.0	1.372	9114.6	911.5
18.00	1.50	99.0	0.776	4529.3	503.3
	2.00	128.0	1.004	5546.7	616.3
	2.50	155.0	1.215	6367.9	707.5
15.00	1.50	81.0	0.635	2490.8	332.1
	2.00	104.0	0.815	2998.7	399.8
	2.50	125.0	0.980	3385.4	451.4
12.00	1.50	63.0	0.494	1181.3	196.9
	2.00	80.0	0.627	1386.7	231.1
	2.50	95.0	0.745	1527.9	254.7

Metric elliptical tube properties

Major axis (mm)	Minor axis (mm)	Thickness, t (mm)	Area, A (mm²)	Weight, w (kg/m)	I minor (mm⁴)	Z minor (mm³)
28	12	1.5	87.2	0.683	1480	246.7
32	15.7	1.5	105.3	0.826	3163	402.9
32	16.7	2	140.4	1.101	4501	539.0
40	16.7	2	165.6	1.298	5525	661.7

Glossary of automotive terms

The first main use of each glossary term is shown in **bold italic** typeface in the text.

Ackermann steering – Steering geometry which causes the inner wheel to turn more than the outer wheel. *100% Ackermann* means that the wheel angles form tangents to the wheel paths and hence no scrubbing takes place.

Anti-dive – Suspension geometry which counteracts the effect of the front-end lowering during braking.

Anti-lift – Suspension geometry which counteracts the effect of the rear-end lifting during braking.

Anti-roll system – Often a torsion bar that connects left and right wheels together in such a way that it stiffens roll but not bump and rebound.

Anti-squat – Suspension geometry which counteracts the effect of the rear-end lowering during braking.

Balanced – The condition of a car that enables the driver to keep a constant steering angle to maintain a given corner radius at increasing speed. Usually this means that the front and rear slip angles are the same. At the limit the car will drift sideways to the outside of the bend.

Bell-crank – A rotating member that connects a pull/pushrod to a spring/damper.

Bump – Suspension movement when the wheels move upwards in relation to the chassis. Also known as **jounce**.

Bump steer – Rotation of wheels about the steering axis during suspension travel.

Camber – In front-view, the angle between the plane of a wheel and the vertical. Positive camber occurs when the top of a wheel is further from the longitudinal centre-line than the bottom.

Camber recovery – Wheel movement during roll that counteracts the development of adverse positive camber in the outer wheel.

Caster angle – In side-view, the inclination of the steering axis relative to the vertical.

Caster trail – In side-view, the distance between the steering axis and a vertical line passing through the centre-line of the wheel when both are projected down to road level.

Glossary of automotive terms

Critical damping – Causes a mass to return to its neutral position without overshoot.

Damper – A device for preventing oscillations of a spring/mass system. Normally based on forcing a viscous fluid through an orifice. Magnitude of force is velocity dependent.

Damping coefficient – The damping force at 1 m/s.

Differential – A mechanism that enables the driven wheels to rotate at different speeds when cornering.

Dive – The tendency of the front suspension to lower as a result of longitudinal load transfer during braking.

Driveshaft – A component that transmits power from the differential to the wheel hub.

Impact attenuator – A structure designed to absorb kinetic energy in the event of a collision.

Induced tyre drag – The component of lateral grip which opposes the forward motion of the car.

Instant centre – The intersection point between the planes of the upper and lower wishbones, about which the wheel rotates. The instant centre moves as the geometry of the wishbones changes.

Jacking – Lift of a car in rebound during cornering.

Jacking down – Gradual lowering of a car because over-damping prevents the recovery of a spring.

Lateral load transfer – The change in wheel loads caused by cornering.

Lateral tyre (or wheel) scrub – Sideways movement of the tyre contact patch during suspension travel.

Longitudinal load transfer – The change in wheel loads caused by acceleration and braking.

Master cylinder – A piston and cylinder component that converts brake pedal force into hydraulic system pressure for braking.

Monocoque – A 3-D structure formed from plates and shells which ideally form a closed box or cylinder.

Oversteer – The condition of a car that requires the driver to decrease the steering angle to maintain a given corner radius at increasing speed. Usually this means that the rear slip angles are greater than the front slip angles and that the rear tyres are closer to their grip limit. At the limit the car leaves the circuit pointing backwards after spinning.

Percentage traction slip ratio – See ***Slip ratio***.

Pneumatic trail – The distance by which the grip resultant force in the tyre contact patch lags behind the wheel centre.

Power – The rate of doing *work* = force × speed or torque × angular velocity.

Progression (gears) – Gear ratios gradually moving closer together in the higher gears.

Pullrod – A tension member that generally connects the outer node of the upper wishbone to the spring bell-crank. It supports the vertical wheel loads.

Pushrod – A compression member that generally connects the outer node of the lower wishbone to the spring bell-crank. It supports the vertical wheel loads.

Rebound – Suspension movement when the wheels move downwards in relation to the chassis. Also known as **droop**.

Ride rate – The combined stiffness of the wheel and tyre.

Rocker-arm – A type of suspension where one wishbone plus the pullrod/pushrod is replaced by a double cantilever beam in bending.

Roll – Suspension movement during cornering when, relative to the chassis, the inner wheel moves down and the outer wheel moves up.

Roll axis – The line joining the roll centres at each end of the car.

Roll centre – The point about which the sprung mass rotates in roll. It can move with increasing roll.

Roll couple – The perpendicular distance between the sprung centre-of-mass and the roll axis.

Roll gradient – Number of degrees of roll per lateral *g* force.

Roll rate – The roll couple to cause one degree of roll.

Rolling resistance – A force causing drag on a free-rolling wheel as a result of energy lost in distorting the tyre.

Scrub radius – In front-view, the distance between the steering axis and the centre-line of the wheel when both are projected down to road level.

Self-aligning torque – The moment produced by the lateral grip force × pneumatic trail.

Slave cylinder – A piston and cylinder component, often in the wheel assembly, that converts hydraulic system pressure into a clamping force on the brake disc.

Slip angle – The angle difference between the plane of a wheel and the direction of motion during cornering. It is caused by distortion of the tyre tread and not skidding.

Slip ratio – Applicable to tyres undergoing acceleration or braking and is the ratio between the rotational speed of the wheel compared to a free-rolling wheel.

Space-frame – A 3-D structure usually formed from tubes and ideally forming triangles.

Spring rate – The stiffness of a suspension spring.

Sprung mass – The mass of car supported by the suspension springs which includes the chassis frame, bodywork, engine and driver. Often half the mass of the suspension links and driveshafts is considered to be sprung mass.

Squat – The tendency of the rear suspension to lower as a result of longitudinal load transfer during acceleration.

Static toe – In plan-view, *toe-in* occurs when the centre-lines of the wheels on one axle converge in front of the car. *Toe-out* is when they diverge. It can be measured in terms of angle or the difference between the distances from the car longitudinal centre-line to points on the front and rear of the wheel rim at axle height.

Steering axis inclination – In front-view, the angle between the wheel upper and lower ball joints. Also known as **kingpin inclination (KPI)**.

Swing arm – The distance between a wheel centre and its instant centre.

Terminal velocity – The maximum speed of a vehicle reached when all the engine power is used up overcoming aerodynamic drag and other losses.

Torque – The moment generated by a rotating force acting at a radius.

Tyre sensitivity – The phenomenon that the grip of a tyre increases with vertical load but at a decreasing rate, i.e. the relationship between vertical load on a tyre and its grip is non-linear.

Understeer – The condition of a car that requires the driver to increase the steering angle to maintain a given corner radius at increasing speed. Usually this means that the front slip angles are greater than the rear slip angles and that the front tyres are closer to their grip limit. At the limit the car leaves the circuit pointing forwards with the steering on full lock after failing to complete the corner.

Unsprung mass – The mass of car supported directly by the road without passing through the suspension springs, which include the wheels and wheel assemblies. Often half the mass of the suspension links and drive-shafts is considered to be unsprung mass.

Wheel centre rate – The stiffness of a wheel relative to the chassis.

Wheel hop – A wheel bouncing over the road surface owing to lack of damping.

References

1. Adams, Herb, *Chassis Engineering*, HP Books, Penguin Group, NY, 1993
2. Aird, Forbes, *The Race Car Chassis*, HP Books, Penguin Group, NY, 2008
3. Avon Tyres Motorsport, http://www.avonmotorsport.com/resource-centre/downloads
4. Bastow, Donald, Howard, Geoffrey and Whitehead, John P., *Car Suspension and Handling*, SAE International, PA, 1993
5. Daniels, Jeffrey, *Handling and Roadholding*, *Car Suspension at Work*, Motor Racing Publications, UK, 1988
6. Deakin, Andrew et al., *The Effect of Chassis Stiffness on Race Car Handling Balance*, SAE Technical Paper 2000-01-3554.
7. Dixon, John C., *Tyres Suspension and Handling*, SAE International, PA, 1996
8. DTAfast, http://www.dtafast.co.uk/
9. Emerald Engine Management Systems, http://www.emeraldm3d.com/
10. Hexcel, http://www.hexcel.com/
11. Katz, Joseph, *Race Car Aerodynamics*, Bentley Publishers, MA, 2006
12. LISA 8.0.0, http://lisafea.com/
13. Lotus Engineering, Norfolk, England, http://www.lotuscars.com/gb/engineering/engineering-software
14. McBeath, Simon, *Competition Car Aerodynamics*, Haynes Publishing Group, UK, 2006
15. Milliken, William F. and Milliken, Douglas L., *Race Car Vehicle Dynamics*, SAE International, PA, 1995
16. Motec Engine Management Systems, http://www.motec.com/
17. Pacejka, Hans B., *Tyre and Vehicle Dynamics*, Butterworth-Heinemann, 3rd edn, 2012
18. Pashley, Tony, *How to Build Motorcycle Engined Racing Cars*, Veloce Publishing, UK, 2008
19. Pawlowski, J., *Vehicle Body Engineering*, Century, 1970
20. Power Commander, http://www.powercommander.com/
21. Puhn, Fred, *How To Make Your Car Handle*, HP Books, Penguin Group, NY, 1981
22. Seward, Derek, *Understanding Structures*, 5th edn, Palgrave Macmillan, 2014
23. Smith, Carroll, *Prepare to Win*, Aero Publishers, Inc., Fallbrook, CA, 1975

References

24. Smith, Carroll, *Tune to Win*, Aero Publishers, Inc., Fallbrook, CA, 1978
25. Staniforth, Allan, *Race and Rally Car Source Book*, Haynes Publishing Group, UK, 1988
26. Stone, Richard, *Introduction to Internal Combustion Engines*, Palgrave Macmillan, 4th edn, 2012
27. SusProg 3D, http://www.susprog.com/
28. UIUC Airfoil Coordinates Database, http://www.ae.illinois.edu/m-selig/ads/coord_database.htm/
29. Van Valkenburgh, Paul, *Race Car Engineering and Mechanics*, published by author, 1992

ETB Instruments Ltd – DigiTools Software, http://www.etbinstruments.com
SolidWorks, http://www.solidworks.co.uk/
Punch!ViaCAD, http://www.punchcad.com/p-27-viacad-2d3d-v9.aspx

Index

Acceleration, 10
   and downforce, 24
   power-limited, 14
   traction-limited, 12
Ackermann steering, 161
Aerodynamic balance, 220
Aerodynamic design approach, 222
Aerodynamic downforce, 24, 201
Aerodynamic drag, 208
Air induction, 228
   forced induction, 232
   normally aspirated, 228
Air properties, 202
Anti-dive, 71
Anti-lift, 73
Anti-roll system, 111
   calculation, 113
Anti-squat, 73
Axle design, 164

Balance of car, 20, 133
   aerodynamic, 220
   calculation of, 147
   factors affecting, 139
Barge board, 219
Bearing design, 166, 189
Bending components, 47
Bernoulli equation, 204
Boundary layer, 207
Brake system, 193
Braking, 16, 193
   and downforce, 26
Bump steer, 163

Camber, 62, 128
Camber recovery, 67
Car set-up, 241
Carbon fibre composite, 44
Caster angle, 157
Caster trail, 157
Centre of mass, 4
Centrifugal force, 19
Centripetal force, 18

Chassis types, 33
Clutch, 178
Compression components, 47
Computer analysis, 52
Cooling system, 237
Corner weights, 242
Cornering, 18
   and downforce, 28
Crash protection, 56

Dampers, 104
   selection, 109
Data acquisition, 246
Design issues, 30
Differential, 182
   supporting, 184
Diffuser, 218
Double wishbone, 63
Downforce, aerodynamic, 24
Drag coefficient, 209
Driveshafts, 186
Dry sump, 239

Elements of racing, 1
Engine management, 234
Engine systems, 227
Exhaust system, 234

Finite element analysis, 52
Floor-pan, 217
Four-stroke cycle, 227
Friction circle, 23
Front view swing arm (fvsa), 65
Front wheel assembly, 154
Front/rear weight balance, 6
Fuel mixture, 236
Fuel system, 233

Gears, 177
g–g diagram, 23
   effect of downforce, 29

Handling curve, 149

Index

Impact attenuator, 56

Jacking, 150

Kingpin inclination (KPI), 156

Laminar flow, 204
Lateral load transfer, 18
   individual wheel loads, 133
Lateral tyre scrub, 68
Launch control, 191
Load cases, 45
Longitudinal load transfer, 10, 16
Lubrication, 238

Momentum, 210
Monocoque, 42
   materials, 43
Motion ratio, 100

Natural frequency, 94
Neutral handling, 122

Oversteer, 20, 120

Pacejka magic formula, 131
   deriving coefficients, 250
Pneumatic trail, 125
   pre-load, 103

Rear wheel assembly, 176
   requirements, 69
Reynold's number, 206
Ride height, 242
Ride rate, 92
Roll axis, 67
Roll centre, 65
Roll couple, 66
Roll rate and gradient, 92

Safety, 245
Safety factors, 45
Scrub radius, 156
Self-aligning torque, 125
Space frame, 36
   design process, 40
   tubes, 42
Springs, 92
   length calculation, 102
   rate, 92
   specifying, 100
Sprung mass, 95
Static wheel loads, 6
Steering, 160
   Ackermann, 161
   bump steer, 163
Steering axis inclination, 156
Strake, 220
Stressed skin, 41
Structural element design, 46
Supercharger, 232
Suspension, 61
Suspension case studies, 86
Suspension set-up, 242
   hard or soft?, 92
   natural frequency, 94

Tension components, 46
Testing, 245
Torsional stiffness, 34
Traction circle, 23
Turbocharger, 232
Turbulent flow, 204
Tyre sensitivity, 3, 22
Tyres, 120
   cornering forces, 124
   induced drag, 125
   modelling, 131
   rolling resistance, 15, 125
   test data, 128
   under acceleration and braking, 126

Underfloor, 217
Understeer, 20, 120
Universal joint, 187
Unsprung mass, 95

Vortex generator, 219

Wheel centre rate, 92
Wheel nuts, 174
Wheel uprights, 169, 190
Wheelbase and track, 31
Wings, 212
   issues, 216
   selection, 215
Wishbone design, 77